Learning Robotics, with Robotics, by Robotics

**Human-Machine Interaction Set**

coordinated by
Jérôme Dinet

Volume 3

# Learning Robotics, with Robotics, by Robotics

*Educational Robotics*

Ilaria Gaudiello
Elisabetta Zibetti

iSTE

WILEY

First published 2016 in Great Britain and the United States by ISTE Ltd and John Wiley & Sons, Inc.



ISTE Ltd
27-37 St George's Road
London SW19 4EU
UK

www.iste.co.uk

John Wiley & Sons, Inc.
111 River Street
Hoboken, NJ 07030
USA

www.wiley.com



Library of Congress Control Number: 2016947850

British Library Cataloguing-in-Publication Data
A CIP record for this book is available from the British Library
ISBN 978-1-78630-099-7

# Contents

# Foreword

There has been a growing interest in the use of educational robots in schools. In the 1960s, after Seymour Paper introduced the LOGO programming language and the floor turtle, robotics became an issue in the educational environment. Robots are slowly being incorporated into our society and currently, the number of service and/or assistance robots has outnumbered industrial robots. So robots are slowly beginning a process of seamless integration into our everyday lives both at home and at school where their applicability is at the core of an increasing number of studies [ALI 13, MUB 13]. However, this does not include "robots for kids": the impact of educational robotics is even more crucial for children and teenagers, where robots can be used for their cognitive development and intellectual growth. As a consequence, greater attention must be paid to how educational robots can be better integrated into the lives and into the education of young people.

Traditionally, the majority of studies investigating "educational robotics" has attracted the interest of teachers and researchers as a valuable tool to develop cognitive and social skills for students from pre-school to high school and to support learning in different domains such as science, mathematics, technology, informatics and other school subjects or interdisciplinary learning activities. Even though a review of the scientific literature reveals that educational robotics is a growing field with the potential to significantly impact the nature of science and technology education at all levels, from kindergarten to university, this book is very original for three reasons:

1) In this book, educational robotics is viewed from a psychological point of view, i.e. from a human-centred approach. For some researchers, the main goal of our project is to understand the current and future needs of the robotics industry, the current robotics curriculum, and to analyze the gap that might occur between the two. In my opinion, the main goal is to understand the current and future needs of the users, the users being learners and teachers;

2) If there are many attitudes and opinions about educational robotics produced without scientific arguments, this book provides serious scientific answers to three questions:

– is educational robotics just a servant of other subjects? No. A wider range of possible robotic applications has the potential to engage young people with a wider range of interests [AMI 12]. Pursuing this challenge we need to develop new and innovative ways to increase the attractiveness and learning benefits of robotics projects. And different strategies exist for engaging a broad range of young learners in robotics [RUS 08]: projects focusing on themes, not just challenges; projects combining art and engineering; projects encouraging storytelling; organizing exhibitions, rather than competitions;

– is educational robotics just a fad? Yes and no. Even if robots can have positive educational benefits, they are no panacea [AMI 12]. In the scientific literature, there have been some studies reporting non-significant impact on learners observed in some cases [BEN 11]. It's the reason why the impact of the Educational Robotics in promoting student learning and in developing sensori-motor and/or cognitive skills needs to be validated through research evidence and scientific proofs. But …

– is educational robotics an excellent tool for teaching? It depends … Empirical and experimental studies are presented.

3) If educational robotics is a broad term that refers to a vast collection of different activities, instructional programs, physical

platforms, educational resources and pedagogical philosophy, this books proposes an innovative distinction between the following approaches associated with educational robotics:

– for "learning robotics", students use a robot as a platform to learn robotics, or, more broadly, engineering (i.e. mechanics, electronics, and programming) in a hands-on and collaborative way;

– for "learning with robotics", robots are used as human-like (e.g. robots such as Nao, Qrio, Rubi, Roobovie, iCub) or animal-like (e.g. robots such as Aibo, Pleo) assistants for teachers (e.g. displaying multimodal content) or companions for pupils and students (e.g. connecting images and words, memorizing new words of a foreign language).

Finally and as Alimisis [ALI 13] said, "*the role of educational robotics should be seen as a tool to foster essential life skills (cognitive and personal development, team working) through which people can develop their potential to use their imagination, to express themselves and make original and valued choices in their lives. Robotics benefits are relevant for all children*".

Jérôme DINET
University of Lorraine
July 2016

# Preface

This book is about how educational robotics (ER) is affecting the representation, acceptance and learning of its users. Through a psychological perspective, the book discusses the three ER learning paradigms that are distinguished by the different hardware, software and correspondent modes of interaction allowed by the robot: (1) *learning robotics*, (2) *learning with robotics* and (3) *learning by robotics* [TEJ 06, GAU 14].

For *learning robotics* [ALI 09], students use a robot as a platform to learn robotics or, more broadly, engineering – i.e. mechanics, electronics and programming – in a hands-on and collaborative way [PET 04, LIU 10, SOA 11, BEN 12]. In this framework, our objective is to investigate learning robotics under the issue of mental representation [GAU 15]. Here, the underlying research question is which representation users hold about robots when constructing and programming a robot? By robot representation, we mean its ontological and pedagogical status and how such status changes when users learn robotics. In order to answer this question, we will present an experimental study that we carried out based on pre- and postinquiries, involving 79 participants. The results show that building and programming a robot (Bot'n Roll®) fosters a more nuanced judgment about robots' belonging to the living and non-living categories but, on the other side, a more definite judgment about the pedagogical roles that a robot may serve.

For *learning with robotics* [DAU 03], robots are used as human-like (e.g. robots such as Nao, Qrio, Rubi, Roobovie and iCub) or animal-like (e.g. robots such as Aibo and Pleo) assistants for teachers – e.g. displaying multimodal content [HYU 08] – or companions for pupils and students – e.g. connecting images and words [TEJ 06], memorizing new words of a foreign language [MOV 09, CHA 10]. In this framework, our objective is to investigate learning with robotics under the issue of users' functional and social acceptance of robot [KAP 05, AVR 13, LE 13, DIN 14, DIN 15, FRI 14, DE 15]. Here, the underlying research questions are: do students trust in robot's functional and social savvy? Is trust in functional savvy a prerequisite for trust in social savvy? Which individuals and contextual factors are more likely to influence this trust? In order to answer these questions, we will present an experimental study we have carried out with 56 participants and an iCub robot [IVA 13, IVA 16]. Here, trust in the robot is considered as a main indicator of acceptance in situations of perceptual and sociocognitive uncertainty and is measured by participants' conformation to answers given by iCub. In particular, we are interested in understanding whether specific user-related features (i.e. desire for control), robot-related features (i.e. attitude toward social influence of robots) and context-related features (i.e. collaborative vs. competitive scenario) impacted trust in iCub. The results show that participants conformed more to iCub answers in functional than in social task. Moreover, the few participants conforming to iCub answers in the social task also conformed less in the functional task: trust in robot's functional savvy was not a prerequisite for trust in social savvy. Finally, desire for control, attitude toward social influence of robots and type of interaction scenario did not have an impact on trust in iCub.

Contrary to these two preceding learning modes that have been labeled as robotic-assisted instruction [VAN 91] – in so far the robot is a passive assistant of the teacher or a passive platform for the students – learning by robotics is named robotic-based instruction (RBI [KIM 14]), in so far the robot constitutes a medium between the students, the school subjects and the teacher: the robot is a tool – i.e. a constructible and programmable kit – that tangibly embodies the concepts of the lesson, and stimulates creative and collaborative problem solving [DEN 94].

For *learning by robotics* [RES 96, PAP 80], students learn both about the content of the lesson and about robots (Lego Mindstorms®, Lego WeDo®, PicoCricket®, Robotami®, etc.), by acquiring subject-specific knowledge [BAR 09a, WHI 07, HUS 06] as well as transversal competences [DEN 01, LIN 07, SUL 08], and fostering the four dimensions of learning – cognitive, affective, social and meta-cognitive [CAT 12]. Although by taking the role of facilitator, the teacher is not seen anymore as the only owner of the knowledge or as the evaluator of students' performance, but he/she catalyzes students' ideas around a concrete activity and guides their progress [GAT 03, SUL 09]. In this framework, our objective is to investigate learning by robotics under the issue of impact of RBI on students' knowledge and competence acquisition (when educational robots are used within a specific pedagogical approach, that is inquiry-based science education (IBSE) [QUI 04, BEL 10, RIE XX, GAT XX]. Here, the underlying research questions are as follows: to what extent the combined RBI and IBSE frame [WIL 07, EGU 12, DEM 12, RIB 12] has a positive impact on cognitive, affective, social and meta-cognitive dimensions of learning? Does this combined educational frame improve both domain-specific and non-domain-specific knowledge and competences of students? In order to answer these questions, we will present an experimental study carried during a 1-year RBI and IBSE in the frame of the RObeeZ school project[1]. The longitudinal experiments that involved 26 pupils and two teachers was based on assessment jointly elaborated by teachers and researchers in order to evaluate the RBI and IBSE effects on four dimensions of learning [FLA 79, SHO 89, VER 96, SAL 98, AND 01] as well as on grades attributed by teachers for evaluating students' knowledge and competences. Main results show significant improvements in mathematics (measures, geometry and problems) and positive impact of RBI and IBSE on the four dimensions of learning.

1 The research has been made through the FP7 EU project Pri-Sci-Net: http://www.prisci.net/.

The recent field of investigation of effects of ER on learning is extensively spreading in scholar and extra-scholar contexts. At the crossroad of artificial intelligence, psychology and science of education, our book discusses how the processes of these learning paradigms (*learning robotics*, *learning with robotics* and *learning by robotics*) might be improved.

> A robot […] is virtually a chimera: all of its components are real, yet it does not exist as an entity. It will affect and transform our lives similarly to the discovery of fire and the inventions of the wheel, the steam engine and the mobile phone. But will it transform *us*? This fascination with robots is merely an expression of humanity's seemingly endless ability for discovery: leaving Africa to go and discover what lay beyond. Arriving in Asia and from there Europe and America. Prehistoric men discovered the American continent through its Northern point by crossing the Bering Straight when it was frozen over; they explored it from one end to the other and only then did they begin to dream. The oldest painted caves of the continent are located to the South, where man had reached the end and had nowhere left to explore, no looking glass to go through, other than through thought. The walls of these caves are covered in carvings of men with animal-faces. Robots represent the last frontier for men who have conquered lands, seas and danced with the stars; the only Universe left to explore is themselves.
>
> M.N HIMBERT (2012). *Le Robot Pensant*, pp. 201.
> Paris: Editions du Moment.

> "Teach me to imagine a result without mourning if it emerges differently"
>
> P. ARTISAN'S

> The *Robolution* isn't a rhetorical term or a marketing strategy. It is an entirely new approach to Science and Technology. This *Robolution* causes so many upsets to our way of life that it is essential to think about it not only in economic terms but in pedagogical terms as well (…) Most robots are so recent that their perceived value is often higher than their real value. (…)
>
> B. BONNEL (2010) *Viva la Robolution*, pp. 279–284. Paris: Editions JCLattès.

Ilaria GAUDIELLO
Elisabetta ZIBETTI
July 2016

# Introduction: Educational Robotics

The process of democratization of technology that has taken place since 1980 in the professional, tuition and entertainment spheres has paved the way for a renewal of education. Soon after the computer entered our society, Papert and Solomon [PAP 72] published "Twenty things to do with a computer". At that time, these authors observed that, when asked what they thought about computers in education, people had different ideas. Some imagined future students as computer programmers: these people thought that the next generation would have learnt and mastered programming as a normal process of alphabetization; others, by contrast, apprehended the possibility that the computers would have "programmed" the students, i.e. a massive use of technology in education could have irreversibly transformed students ways of thinking and communicating in a machine-like manner.

Today, a new technological revolution has started, namely the robolution [BON 10]. This revolution seems to be so powerful and pervasive that our times have been defined as "the era of the robot". Daily use of robotics is encouraged in an extensive range of domains, among which is the educational domain. However, caution should be used with regard to a revolution that could be dictated by industrial development and technological progress more than by authentic educational needs.

It thus becomes urgent to understand the usefulness of integrating robots in the educational system. Such urgency results in the emergence of a new specific field of study: educational robotics (ER) [EGU 10]. ER aims to introduce to the classroom a variety of embodied artificial intelligence technologies (human-like as well as animal-like robots and robotic kits). According to Bussi and Mariotti [BUS 09, p. 2], who borrow from Vygotsky's notion of semiotic mediation [VYG 78], educational robots are intended as "semiotic tools":

> "(…) semiotic potential resides in any artifact consisting of the double semiotic link that the artifact has with both the personal meanings that emerge from its use and the knowledge evoked by that use (…) in educational settings".

By means of such tools, the general objective of ER is to scaffold and renew teaching on the one side and learning on the other side [DEN 94].

After 30 years since the arrival of Logo Turtle[1] [PAP 80, PEA 83, KLA 88, CLE 93], the first educational robot, we believe it is time to clarify the nature of ER and to start thinking about "Twenty things to do with a robot", in particular with an educational robot – Appendix 1 [RES 96].

In order to do this, we will first outline the historical origins of ER and describe its position with respect to other current information and communication technologies (ICTs). Then, we will illustrate the three learning paradigms presently supported by the types of robots available on the market: *learning robotics*, *learning with robotics* and *learning by robotics*. These three learning paradigms are the focus of our research and motivate the tripartite structure of this book. Their definition is of pivotal importance for introducing our three experimental investigations and will therefore be deepened all along the present work. Finally, we will present the research questions from which we have moved to develop this work.

## I.1. Origins, positioning and pedagogical exploitations of ER

ER finds its origins in a historical moment where the gap dividing the generation of "digital natives" and the previous one of "digital

1 http://turtleacademy.com/.

immigrants" becomes manifest in terms of technology fluency and ways of thinking [PRE 01]. Surrounded by digital technologies from their birth, young people today might treat information differently from their predecessors, who nowadays experience difficulties in adapting to such an omnipresence of technology.

If so, this technogenerational gap is particularly relevant in educational contexts, where these two generations, represented by teachers and students, interact to develop new knowledge and competences by using educational tools, which are capable of shaping students' intellectual growth. For this reason, a debate has been raised about limiting new technologies to extra-school contexts (e.g. summer campus and competitions) versus employing them at school [ARR 03]. Although education is already familiar with questions about the suitability of technologies in the class, it is indeed new to questions about the suitability of this specific technology, i.e. robotics. It is thus crucial to systematize theoretical and experimental knowledge about ER to understand its possible applications and consequent impacts on education. In fact, though being still a "babbling" discipline [MAT 04], ER already presents three fundamental characteristics: (1) a multidiscipline heritage, (2) a specific positioning with respect to other current ICT, and (3) different hardware–software combinations, which serve different pedagogical exploitations. In the following sections, we will examine these three characteristics to delineate the identity of ER.

## I.2. A cross-disciplinary heritage

ER is at the crossroads of three disciplines belonging to the broader area of research of cognitive sciences: psychology, educational science and artificial intelligence.

Fundamental studies from psychology on the role of experiential learning [PIA 52], intrinsic versus extrinsic motivation [LEP 00], social dynamics of learning [VYG 78] and meta-cognition [GAG 09] are crucial for investigating the mental processes implied by the use of a new technology for educational objectives [AND 08].

Educational sciences, which seek to implement research on cognitive and emotional mechanisms at play during learning

[MEL 09], provide a number of case studies that are representative of current pedagogical approaches [BRU 02], monitor trends in learning results – see, for example, OECD-PISA (The Organisation for Economic Cooperation and Development-Programme for International Student Assessment)[2] – and also support the design of guidelines for the adaptation of the educational system to contemporary society [VOS 01].

Artificial intelligence [HEU 94], more recently labeled as "cognitive informatics" [WAN 10], continuously raises new challenges in terms of robot prototypes with physical and functional features engendering a variety of interaction possibilities. In this sense, ER confronts young students with a technology at the boundary of living and non-living entities, which can be built and programmed for obtaining specific functions and behaviors [MAR 00].

We argue that it is the combination of these three disciplines that contributes to defining the technological status and pedagogical exploitations of educational robots, as distinct from previous educational technologies.

## I.3. The educational robot: an ICT like others?

In the last 20 years, different types of technologies, suited for different educational exploitations, have appeared. A variety of educational softwares have been conceived for interactive learning on traditional hardware supports (computers, tablets, etc.) [DE 01]. Other tools – such as the e-learning platforms [ROS 01] and the digital schoolbag [TIJ 06] – allow customization of the educational interface according to students' needs.

Critical reflections about the integration of ICT at school have been at the heart of committed debates among educators, researchers and decisions makers, engendering questions such as "What is the role of media in education?" and, among the media, "What is the role of the

2 http://www.oecd.org/fr/edu/scolaire/programmeinternationalpourlesuivideacquisde selevespisa/.

computer?". With the birth of ER, further questions have been raised: what similarities do robots share with their technological precursors? What distinguishes the former from the latter?

As a first answer, two features of the robot and of its precursor, the computer, can be examined: their "technological status" and their "pedagogical exploitations".

With respect to *technological status*, the computer presents a double specificity: this technology can be either an end in itself – i.e. an engineering object that it is employed as a platform to understand how computers are assembled and programmed – or an ICT [AND 08] that can be defined as a medium, a processor and a tool [BAR 96]. As a "medium", the computer supports software that students use to interactively acquire new knowledge. As a "processor", the computer facilitates treatment and storage of information in a way that is specific to the type of content. As a "tool", the computer can be employed to elaborate documents, visualize numerical data, etc.

If we apply this distinction to robots, we find that, as an ICT, the robot can be defined in terms of object – i.e. a constructible and programmable device that can be used to learn mechanics, electronics and informatics (e.g. [MIK 06]) – or of tool – i.e. a device employed to acquire new knowledge and competences [ION 07].

With respect to *pedagogical exploitations*, when using a computer, students can learn either "from" or "with". In the first case, the computer is used to augment pupils' knowledge with software, which facilitates the understanding of subject-related knowledge [BOT 02]. In the second case, technology can be applied to enhance higher-order thinking skills [RIN 02]. This is the case of those software that aim at developing meta-cognitive competences [ZIB 11], as well as motivation and engagement [PRE 05].

Although the pedagogical exploitations of educational robots are related not only to the type of software but also to the type of hardware, as embodied artificial intelligence entities, robots are endowed with shapes and behaviors that add something to computers and consequently raise new learning paradigms.

## I.4. Three learning paradigms of ER

In the past 30 years, the extensive spread of educational robots in scholarly and extra-scholarly contexts has led to emergence of three ER learning paradigms, distinguished by the different hardware, software and corresponding modes of interaction allowed by the robot: *learning robotics*, *with robotics* and *by robotics* ([TEJ 06, GAU 14], see Table 1.1).

| | Type of robot | Function/role of the robot | Status | Learning paradigm | Educational objectives |
|---|---|---|---|---|---|
| Robotics-assisted instruction (RAI) | Modular robots or robotics kits (e.g. LEGO WeDo® and Mindstorms®, Bioloid®, Bot'n Roll®, Darwin® and Thymio®) | Platform of construction and programming | Object, end in itself (transparent technology) | *Learning robotics* | Acquiring knowledge and competences in mechanics, electronics and informatics<br>Developing problem-solving and collaborative attitude |
| | Human-like robots (e.g. Nao®, Qrio®, Rubi® and Roobovie®) | Support the teacher and illustrate the lesson; colearn with students | Assistant; peer (black box or semitransparent technology) | *Learning with robotics* | Acquiring domain-specific knowledge<br>(e.g. foreign language) in an interactive<br>multimodal way |
| | Animal-like robots (e.g. Aibo, Furby, Paro and Pleo) | Accompany students in a playful discovery of technology, colearn with them; supporting students with cognitive and emotional impairments | Companion (black box technology) | *Learning with robotics* | Discovering technology in a playful way and acquiring domain-specific knowledge through interaction |
| Robotics-based instruction (RBI) | Robotic kits (e.g. LEGO WeDo® and Mindstorms®, PicoCricket®, Roamer® and Robotami®) | Tool supporting short- and long-term collaborative projects or inquiries on scientific as well as humanistic issues | Tool (transparent technology) | *Learning by robotics* | Acquiring subject-related as well as transversal knowledge and competences, by fostering the cognitive, affective, social and meta-cognitive aspects of learning |

**Table I.1.** *Types and functions of robots along with their technological status, learning paradigms and educational objectives in robotics-assisted instruction (RAI) and robotics-based instruction (RBI)*

In the first learning paradigm (Figure I.1), the robot is an end in itself: students use it as a platform to learn robotics or, more broadly, engineering – i.e. mechanics, electronics and informatics – in a hands-on and collaborative way [MIO 96, AHL 03, PET 04, ROG 04, MIT 08, ALI 09, NUG 10, SUL 15].

**Figure I.1.** *Some examples of robots employed in the educational paradigm learning robotics. From left to right: Bot'n Roll®, Bioloid®, Thymio® and Darwin®*

In the second learning paradigm (Figure 1.2), robots are human-like (e.g. Nao, Qrio, Rubi and Roobovie) or animal-like (e.g. Aibo, Furby and Pleo) assistants, peers or companions that are supposed either to help teachers – for example displaying multimodal content [HYU 08] – or to learn at the same time as the pupil – e.g. connecting words and images [TEJ 06], memorizing new words of a foreign language [KAN 04a, HAN 09, MOV 09, CHA 10], or doing sport [TAN 05]. These two typologies of robots have been defined as "non-transparent" or "black box" [KYN 08] since in most cases their functions or behaviors, which are in-built, cannot be modified: users do not have access to the internal system and cannot reprogram the robot.

In the third learning paradigm (Figure 1.3), the robot is a "transparent" device – i.e. a constructible and programmable kit – that embodies the philosophy of "playing pianos, not stereos" [RES 96]: these robots stimulate students to become authors of educational technology, rather than simply users. By constructing and programming robots, students encounter problems [DEN 94], create projects [ALI 09] and carry out inquiries [DEM 12, EGU 12].

**Figure I.2.** *Some examples of robots employed in the paradigm learning with robotics. From left to right: Qrio®, Rubi®, Roobovie®, Nao®, Aibo®, Paro® and Pleo®*

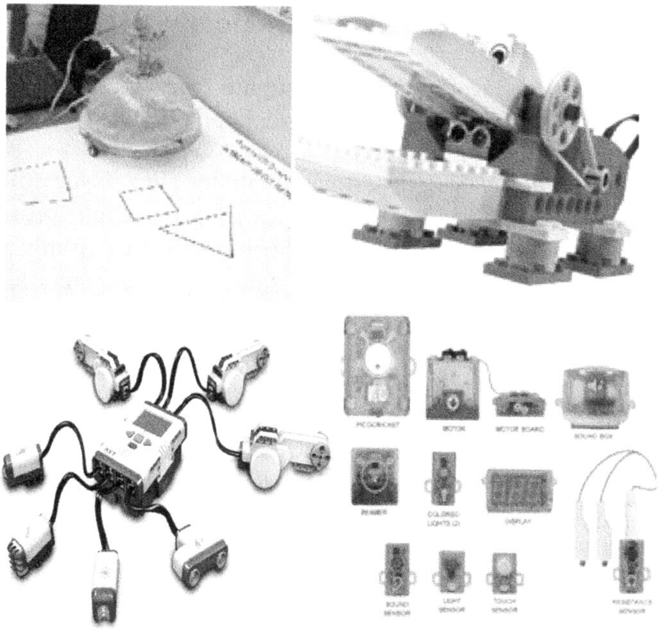

**Figure I.3.** *Some examples of robots employed in the paradigm learning by robotics. From left to right: Logo Turtle®, LEGO WeDo®, LEGO Mindstorms® and PicoCricket®*

While the first and the second learning modes have been labeled as robotic-assisted instruction [VAN 91] – in so far as the robot is a passive assistant of the teacher or a passive platform for the students – the third is called robotic-based instruction (RBI [KIM 14]), in which the robot constitutes a medium among the students, the school subjects and the teacher: students learn both about the content of the lesson and about robots by proposing ideas and solutions, collaborating, relying on the immediate feedback of the robot to evaluate what they do and developing learning strategies, whereas by taking the role of facilitator, the teacher is not seen anymore as the only bearer of the knowledge or as the evaluator of students performances, but he/she catalyzes students ideas around a concrete activity and guides their progress.

## I.5. Research intentions and scientific questions

In the present work, we investigate the three ER learning paradigms from a cognitive psychological perspective.

*Learning robotics* is investigated under the issue of mental representation (Chapter 1). This first learning paradigm presents two specificities: not only are robots at the boundary of living and non-living entities, but students are at the same time designers and users, since they construct and program robots in order to interact with them afterward. Given such a peculiar status of robots and of students in this learning paradigm, and to purposefully prospect and implement applications of robots in our educational system, we argue that it is crucial to approach fundamental issues such as (1) what is the position of robots in students' common-sense ontology? and (2) What pedagogical role(s) should students attribute to robots?

*Learning with robotics* is investigated under the issue of users' acceptance of robots in functional and social tasks (Chapter 2). In this learning paradigm students share tasks with robots and co-learn with it. In order to do this, students need to accept the robot as a functional and social agent. Here, the underlying questions of research are as follows: (1) do students trust in the robot as a functional and social agent? (2) is trust in the robot as a functional agent a pre-requisite for trust in the robot as a social agent? and (3) which individual and contextual factors are likely to influence this trust?

*Learning by robotics* (Chapter 3) is investigated under the issue of the impact of RBI on learning when educational robots are used within a specific pedagogical approach, that is, inquiry-based science education (IBSE). Here, the research questions are as follows: (1) to what extent does the combined RBI and IBSE frame have a positive impact on cognitive, affective, social and meta-cognitive dimensions of learning? and (2) does this combined educational frame improve both students domain-specific and non-domain-specific knowledge and competencies?

In order to answer these questions, we have carried three experimental studies (sections 1.3, 2.3 and 3.3).

In the following chapters, a detailed state of the art on the three learning paradigms will be accompanied by those critical reflections that have elicited the above-mentioned research intentions and scientific questions, constituting a theoretical base to design our experiments:

> "We are particularly interested in how children think about the artificial creatures they build. Do they see them more as machines or as creatures? To what extent do they attribute intentionality to the creatures/machines? It seems that people tend to view creatures on many different levels. Sometimes they view the creatures on a mechanistic level, examining how one LEGO® piece makes another move. Then, they might shift to the information level, exploring how information flows from one electronic brick to another. At other times, people view the creatures on a psychological level, attributing intentionality or personality to the creatures. One creature 'wants' to get to the light. Another creature 'likes' the dark. A third is 'scared' of loud noises".
>
> RESNICK M., WILENSKY U., 1993, p. 69.

> "Perhaps in the future, as robots become more commonplace and children have more opportunities for interaction, their schemata of the living world may need to accommodate (reorganize) to appreciate how robots are similar and dissimilar to humans".
>
> BERAN T.N. *et al.*, 2011, p. 547.

> "We ask not just about where we stand in nature, but about where we stand in the world of artifact. We search for a link between who we are and what we have made, between who we are and what we might create, between who we are and what, through our intimacy with our own creations, we might become. (…) [TUR 05, p. 18]" In the classic children's story 'The Velveteen Rabbit', a stuffed animal becomes 'real' because of a child's love.

Tamagotchis do not wait passively but demand attention and claim that without it they will not survive. With this aggressive demand for care, the question of biological aliveness almost falls away. We love what we nurture; if a Tamagotchi makes you love it, and you feel it loves you in return, it is alive enough to be a creature".

TURKLE S., 2011.

# 1

# Learning Robotics: Users' Representation of Robots

## 1.1. Introduction: the ontological and pedagogical status of robots

In recent years, there has been a growing interest in users' representations of robots within several complementary fields of study: cognitive psychology [KAH 06, JIP 07, BER 08, BER 11], science and technology education [e.g. SLA 11] and anthropology [GRI 12]. The reason for this interest lies in a shared wonder at a new technology which, despite being a manmade entity, i.e. an artifact, has enough power to surpass people in the accomplishment of several physical and decision-making tasks. This mixed definition of the robot, as an entity that possesses at the same time something that is greater than and something that is less than living and non-living beings, seems to challenge traditional ontological categories [SEV 10]. The difficulty involved in assigning robots either to the category of living entities or to that of non-living entities has led researchers from different fields not only to postulate the creation of a completely new category of objects but also to revise the traditional concept of "being alive" itself. In the words of MacDorman and colleagues [MAC 09, p. 486]:

> "Among all human artifacts, perhaps robots share the most in common with their maker. Like computers, and in fact because they are controlled by computers, they

> can process huge amounts of information. Like powered equipment, they can manipulate their environment and move within it. And like dolls, mannequins and other effigies, they can resemble us – either abstractly or down to the dimples of our cheeks. Nevertheless, the differences between machine and maker are profound. Metabolism, life span, sexual reproduction, ancestry, culture and consciousness for now distinguish us from robots. Thus, the similarities and differences between us and them circumscribe a chasm that is at once narrow and deep".

Robotics kits, in particular, have an interesting status. On the one hand, they are engineering objects that Slangen *et al.* [SLA 11] describe as "a system, that is, any group of interrelated parts designed collectively to achieve a designed goal. The system maintains its fundamental structure notwithstanding the possibility of infinite transformations. Systems have input, processes, and output. In order to perform a task a robot integrates solutions to sub-problems from different technological domains (e.g. mechanics, electronics, pneumatics, calculation) into one machine. The robot is a construction and consists of a frame with static components (bricks, pins, beams), dynamic mechanical components (gears, axles), electronic components (sensors, display, bulbs) and electro-mechanical components (motors). Robots should be well designed and constructed with the right components, and be stable and strong enough to enable the execution of the function(s). This requires understanding of concepts like stability, sturdiness, motion, etc. (…) The robot is controlled by means of software designed to enable the robot to function. (…) The performance of a robot is based on the three basic capabilities of sensing (S), reasoning (R) and acting (A), which repeat in succession and form the so-called S-R-A loop [VAN 06]. A robot that is able to sense, reason and act needs hardware components like sensors, a PLC, and actuators (motors, bulbs, speakers, displays)".

On the other hand, there is something unique about robotics kits. To explain this uniqueness, we refer to the work of Severson and Carlson [SEV 10], who first applied the expression "creative control" to robots, meaning that users of this kind of robot experience a form of simulation, projection and personification similar to that experienced by children during imaginative play. That is, the fact that this kind of robot has to be physically created from scratch through assembly procedures and progressively "tamed" through algorithms, implies that users are simultaneously engineers and interpreters of robots' behavior [ACK 91]. This, we believe, requires an imaginative effort consisting of shaping a representation and reshaping it through time, as the robot is developed by its users.

## 1.2. What do we mean by robot representation?

Representations have a pivotal role in cognitive psychology research. They are generally described as mental constructs that may concern the physical world, but also social and mental entities [LE 05]. They can be permanent or occasional, that is stable or triggered by some specific activity or context [STE 09]. Within the robotics literature, the term "representation" often occurs as an umbrella concept that incorporates different meanings. These meanings have been investigated using methods that are traditionally employed to find out how users perceive innovative technologies, but, interestingly, also to examine the status of imaginary or fictional entities such as made-up companions or the characters of cartoon movies [SEV 10], as well as of strangers or out-group members [KUC 12].

In this context, we use the term "representation" of an educational robot to mean (1) the place teenagers assign to robots within their common-sense ontology and (2) the different pedagogical roles (i.e. an object to be constructed and programmed, a tool to learn school subjects, a classmate) they attribute to such robots.

### 1.2.1. *The place of robots in our common-sense ontology*

Each society marks in its own way the boundaries between the categories of beings (living or non-living, real or imagined). These boundaries depend on the features that we use to attribute or deny to the entities that surround us. In general, such features form a system within what is traditionally called ontology [DES 10].

Recent studies, which have addressed the issue of whether robots are considered closer to inanimate objects than to living entities, have demonstrated that robots are considered to be "kind of alive" [TUR 11]. This would imply that being alive involves degrees instead of being a matter of all-or-nothing. More precisely, the literature indicates a considerable variability in young people's ability to classify robots. Robots are sometimes seen as being characterized by features of both living and non-living entities, but at other times depicted as only having machine-like features [OKI 06, JIP 07, SAY 10]. Although disparate, these features can be classified according to three main types: biological, technological and psychological. When we examine children's judgments about robots, such features often occur in couples, particularly in antithetical couples. For instance, children from 3 to 5 years old say that robots are non-living but real [JIP 07], non-living but aware [TUR 11, p. 62], and mechanical but intelligent and capable of emotions [BER 11]. Similarly, adults show inconsistent or paradoxical judgments: they see robots as machines, but "alive enough" to substitute people when they lack the ability to do something (e.g. the ability to listen and understand others) [TUR 11]. Another interesting study performed at the Museum of Quay Branly by Vidal and Gaussier [VID 14] witnesses that when interacting with a robot people show two attitudes that are apparently in contradiction: on the one hand, they try to understand the mechanical functioning of the robot; on the other hand, they interact with the robot as it was a real person. In summary, it seems that neither children nor adults feel comfortable assigning robots solely to either the category of living entities or the category of non-living entities.

### 1.2.2. *Categories: essentialist versus graded*

A category is traditionally defined as set of items which are built upon the broad and defined differences between those items [GEL 03]. According to the essentialist school, an entity contains core features that allow people to decide if it belongs to a certain category or not [GEL 04]; for example, we may decide that an entity is a bird because it flies. As Gelman [GEL 03] remarks, essentialism is a "pervasive, persistent reasoning bias, that affects human categorization in profound ways, a sort of cognitive predisposition that emerges early in childhood, particularly for understanding the natural world" (p. 6). According to Gelman, it seems that people unconsciously believe in hidden essential qualities that are responsible for the observable similarities between different members within a category. In contrast, antiessentialism states that there are no defined essential features of objects, no sufficient and necessary characteristics, and that category membership is not a matter of all or nothing, but is instead graded; for example, a penguin might be judged to be partly a bird and partly a fish, but not completely, because we do not recognize all the features of a bird in this animal, namely the ability to fly rather than swim [KAL 95].

The difficulty that adults, as well as children, experience in trying to build a coherent representation of robots could mean that neither children nor adults adhere to an essentialist view of ontological categories when dealing with the entity "robot". Strikingly, robots seem to threaten the long-established essentialist way of interpreting the world.

In cognitive psychology, the antiessentialist position is explained through models like that of graded representations [MUN 01], according to which representations are not an all-or-nothing phenomenon, but are graded. The strength of a representation would then depend on the contextual cues (e.g. the societal environments) that favor it. That is, being able to assign an entity neatly to a specific category (essentialist) or in a graded way (antiessentialist) is not simply a matter of personal preference, but rather the product of a contextualized common-sense ontology.

The literature shows that if living entities (e.g. animals, plants) and social status (e.g. gender, nationality) are mostly categorized in an essentialist manner, at least in western cultures, the same does not hold for artifacts. Artifacts do not have an essence [SLO 03]; they are a composite set of elements that are continuously replaced and renewed, and they mainly serve a particular function. This contributes to the difficulty we face when we are asked to categorize artifacts like robots. This is such to the point that, according to several authors [KAH 04, SEV 10], robots represent the emergence of a new ontological category (NOC) that is neither alive nor not alive, but something altogether different: one of personified or behavioral technologies.

### 1.2.3. *The NOC hypothesis*

The NOC hypothesis [KAH 12] states that an NOC is emerging, a category that does not map onto humans, animals or artifacts. This means that although natural and artifactual categories have remained relatively stable for tens of thousands of years – so that it has been possible since the origins of modern psychology to study how children develop their categories of the physical and social world – in recent decades, the rate of technological change has increased so rapidly that children's cognition is now constantly in flux and will continue to be so [KAH 12].

Kahn and colleagues believe that this incoming NOC will become more identifiable as other embodied social computational systems (e.g. personified "smart" phones, cars and homes of the future) become increasingly advanced and pervasive.

A similar view is proposed by Turkle [TUR 11] who points out that robots are different, both from living entities and from non-living functional objects or toys, for the reason that people cannot simply project their beliefs onto them. In this respect, robots seem to break the traditional subject–object opposition [KAP 05, p. 142]: they are interactive, they can develop under our guidance and they have a memory. Severson and Carlson [SEV 10] provide five criteria to recognize whether robots constitute an NOC. The first and most relevant of these criteria is that attributions to robots must cut across prototypic categories (e.g. alive and not alive).

Another relevant finding within the literature that we have considered for this study is that gaining experience with robots leads to more nuanced [VAN 96] and species-specific views of robots [BER 08a]. After interacting with a robot, both adults and children seem to treat it as an intelligent entity, but intelligent in a unique way, which is different to the way that living or non-living entities are intelligent. A convincing explanation for the "species specificity" and "uniqueness" that characterizes our mental images of robots, once we have become acquainted with them, is given by Kahn *et al.* [KAH 11]: "just as we perceive the color orange as a unique color, and not merely as a combination of red and yellow, once we become familiar with robots we will see, conceptualize, and interact with them as a unified entity, and not merely a combinatorial set of constituent properties".

### 1.2.4. *Shifting between the different pedagogical roles of a robot*

According to Kaplan [KAP 05], it is because young people are able to perceive the same robot either as a peer, a construction game, or a domestic animal that they are naturally directed to construct their own idea of what these creatures really are. Turkle [TUR 11, p. 62] observed that after the interactions with an AIBO robotic dog, when interviewed, the participants (aged 17 years) revealed that they saw AIBO both as a creature and a machine. In addition, being involved with the inner technical details of the robot did not diminish their attachment to it.

Concerning robots used to learn robotics, as outlined by Severson and Carlson [SEV 10], young people's ability to shift from one role to another is based on a special form of imagination that children develop from 3 to 12 years of age through play. Such a form of imagination implies that any object can acquire a specific status during playtime and become an inert object again once playtime is over [FLA 87]. The shift among the different roles of the robot is precisely what intrigues young people [KAP 05, p. 158]: they generate "multiple parallel representations" [KAP 05, p. 159] or "simultaneous visions" [TUR 11, p. 62] and they behave accordingly to the type of interaction at play. Robotics platforms seem to evoke, in this sense, the very essence of play.

### 1.2.5. *How do we investigate robot representations and the impact of learning robotics on these representations?*

Several studies have addressed the representation of a robot as the concept or the mental image of a robot that people hold based on their familiarity with imaginary or real robots. This kind of representation has mostly been investigated using explicit measures, that is, through closed-question surveys, mostly using a scale format. For example, the Negative Attitude toward Robots Scale [NOM 06] evaluated adults' negative attitudes toward robots by asking respondents to rate first-person sentences expressing situations of interaction with robots (e.g. "I would feel uneasy if I was given a job where I had to use robots"), the social influence of robots (e.g. "I feel that if I depend on robots too much, something bad might happen"), and emotions in interactions with robots (e.g. "I feel comforted being with robots that have emotions"). The Human Likeness Questionnaire [HO 10], again tested with adults, addressed the "humanness" of robots through semantic differential ratings (e.g. "mortal versus without definite lifespan"). The PERNOD (PERception to humaNOiD [KAM 14]) included items such as admiration for technology (e.g. "I could open my heart to this robot"), utility (e.g. "This robot seems to be able to perform only structured routines") and familiarity (e.g. "I feel an affinity toward this robot"). The Interpersonal Attraction Scale [VEE 11] evaluated friendship, bonding, physical proximity and care for robots among children aged 6–10 years. The Social Credibility Scale [JOO 13] examined the social appearance of robots during interactions (good natured versus irritable, cheerful versus gloomy, dishonest versus honest, etc.).

Two methodological aspects of studies using such instruments to examine people's representations of robots deserve discussion. The first aspect concerns the experimental materials used as inputs: participants were dealing with specific kinds of robots, i.e. robots with specific appearances and behaviors. As a result, their representation was irrevocably biased by the specific kind of robot rather than on their purported representations of robots, preventing people from placing it inside or outside a category of entities – since the robot was already presented as belonging to its own category (e.g. in the study by Berstein and Crowley [BER 08a], pictures of robots are labeled as "robot"). Moreover, in these studies, the robots were either artificial animals or humanoids but in all cases they were all prebuilt devices, black box type technologies preprogrammed to react with a definite behavior [KYN 08]; that is, robots used in these studies were not constructed or programmed by participants. The second aspect is that these studies rarely took into account the influence of non-scientific information (science fiction, advertisements, etc.) or previous exposure to robots or people's ideas about them. Nonetheless, prior robot exposure is an important factor. Bernstein and Crowley [BER 08a] showed that children with little prior experience tended to group the robots with familiar objects of similar living status (i.e. various kinds of animals). In contrast, children with greater prior experience attributed a unique pattern of intellectual and psychological characteristics to the robots (e.g. robots were as smart as a cat, but were less psychological) [BER 08a].

## 1.3. Study 1: Robot representation

### 1.3.1. *Aims and rationale*

In order to understand how students shape representations of robots when learning robotics, we designed a pre- and post-questionnaire addressing the following three issues:

1) the place assigned to robots among living and non-living entities;

2) the educational role(s) attributed to robots;

3) whether building and programming a robot has an impact on (1) and (2).

The two questionnaires (pre- and post-) were presented to teenage students participating in a 3-day robotics event (RoboParty®[1]), where they had to build and program a Bot'n Roll® kit. In an attempt to overcome the methodological limitations of previous studies on robots' representations (section 1.2.5), our questionnaire did not

1 http://www.roboparty.org/.

present a specific kind of robot to participants. This precautionary measure was taken intentionally to avoid biasing participants' answers: we did not wish to evoke a particular typology of robot in the students' minds while investigating their representation, as this might influence their responses. Instead, we asked participants to carry out a picture-rating task and a sentence-rating task.

In the first task, participants look at a set of 10 pictures and to answer the question "how much do these images make you think about a robot?" by providing a score on a 1–5 point Likert scale, where 1 indicated "not at all" and 5 indicated "a lot".

In the second task, they had to answer the question "In the classroom, would you prefer: (1) building and programming the robot, (2) using the robot as a tool to learn school subjects, (3) learning with the robot as a classmate", again by providing a score to each of the three sentences on a 1–5 Likert scale. The answer to each question of the questionnaires was considered graded, if the score provided was different from 1 or 5 on the Likert scale.

A higher graded score reflected a non-essentialist categorization, that is a more nuanced assignment of robots to specific ontological categories (living or non-living) and to specific educational roles (object to build and program, tool to learn school subjects and classmate); in contrast, a lower graded score reflected an essentialist categorization, that is a more defined assignment to these ontological categories and educational roles. The questionnaire was filled in twice, before and after building and programming a Bot'n Roll® kit.

### 1.3.2. *Hypotheses*

HYPOTHESIS 1.1.– Students have beliefs about what a robot is and which educational roles it may serve. However, these beliefs can be stable or triggered by some specific activity or context [KAL 95, STE 09] and are probably influenced by previous exposure to robots [BER 08a]. That is, participants hold a representation of robots, in terms of ontological and educational status. Such representation can be more or less graded [MUN 01], which means that robots can be seen as more or less belonging to the categories of living or non-living entities and as serving specific educational role(s). This first hypothesis has engendered the following predictions:

PREDICTION 1.1.– In the picture-rating task of the pre-questionnaire participants should mostly assign a low graded score. That is, we expected to find a global predominance of "1" and "5" on the Likert scale as an answer the question "How much do these images make you think about a robot?". Moreover, we also expected to find a difference between participants who were already familiar with robots and participants who were not: the former should opt for a stronger essentialist categorization (low graded score), than the latter (medium graded score). Thus, according to this prediction, the less familiar participants are with robots, the more they will show an essentialist categorization of robots, that is an all-or-nothing judgment about robots' belonging to living or non-living categories.

PREDICTION 1.2.– In the sentence-rating task, participants should mostly assign a low graded score. That is, we expected to find a global predominance of "1" and "5" on the Likert scale as an answer the question "Would you like your robot to be an object to be constructed and programmed, a learning tool, or a classmates?". Specifically, the mean graded score would be low for participants who were not familiar with robots and medium for participants who were already familiar with robots. Thus, according to this prediction, the less familar participants are with robots, the more they will tend to neatly attribute one specific role to a robot, that is an all-or-nothing judgment about robots' educational status.

HYPOTHESIS 1.2.– The effect of building and programming a robot on common-sense ontological categorization (essentialist versus graded). Consistent with work by Van Duuren and Scaife [VAN 96], Bernstein *et al.* [BER 08a] and Kahn *et al.* [KAH 11], we hypothesized that, after assembling and programming a robot, participants should show a more nuanced and refined view of a robot; that is a robot's association with a specific ontological category should no longer be a question of all-or-nothing (essentialism), it should instead be a question of degree (non-essentialism). Hence, we formulated the following predictions:

PREDICTION 1.3.– The mean graded score for ontological categorization would be generally higher in the post-questionnaire than in the pre-questionnaire.

PREDICTION 1.4.– Participants would tend to revise their general ontological categorization by demonstrating a dichotomous living /non-living categorization in the pre-questionnaire and a tripartite living/life-like/non-living categorization in the post-questionnaire. Thus, we expected that the distribution of prequestionnaire scores would suggest a categorization of pictures in two groups (living/non-living), while three groups (living/non-living/new category) would be identified in the post-questionnaire.

For predictions 1.3 and 1.4, there should be less variation in the score distributions before and after robot programming for participants who were already familiar with robotics kits prior to their participation in RoboParty®.

HYPOTHESIS 1.3.– The effect of building and programming a robot on attribution of educational role(s) to it. Consistent with work done by Kaplan [KAP 05] and Severson and Carlson [SEV 10], we hypothesized that, after the robot-making experience, participants would ascribe different educational roles to the robot, and this to different degrees. Therefore, we made the following prediction:

PREDICTION 1.5.– The mean "graded score" for educational roles would be greater in the post-questionnaire. Again, there should be less variation in score distributions before and after robot programming for participants who were already familiar with robotics kits prior to their participation in RoboParty®.

### 1.3.3. *Method*

#### 1.3.3.1. *The RoboParty® context*

RoboParty® is a 3-day robotics event organized and facilitated by lecturers and volunteers from the University of Minho. The participating teams are invited to create a robot from scratch using only paper and video instructions, with the support of the University staff. The first day of the event is dedicated to building the robot (soldering electronic components, assembling the mechanical parts and personalizing the robot by reproducing or inventing a fictional character). The second day is dedicated to programming the robot (*Basic* for *Picaxe*) and the third day to the testing of the robot, by taking part in three competitive arenas (*rescue, dance* and *chase*). During the three days, participants can also benefit from lectures on robotics by invited speakers, sports and various leisure activities. The objective for the event is to learn mechanics, electronics and programming through an entertaining and hands-on approach [SOR 11]. A special focus is thus given to pedagogy, collaboration and creativity, in contrast with other robotics events that promote competition instead (e.g. First and Junior LEGO® leagues and RoboCup).

#### 1.3.3.2. *Participants*

111 teams participated in RoboParty®, with each team including a maximum of four participants and an educator; a total of 400 participants were involved, all from urban and rural state schools in Portugal, and from a wide variety of socioeconomic environments. These students were particularly appropriate as a sample population for the purposes of this investigation because, due to the nature of their scientific and technical studies, they already possessed generic knowledge about computer science. Among the 400 RoboParty® participants, 226 volunteered for this study. From the 226 (pre-) and 197 (post-) individual questionnaires collected, only 89 participants had correctly filled in both the pre- and the post-questionnaire. This was due to the fact that, although team leaders had been correctly briefed and the staff had ensured the proper delivery of instructions, data were collected during a robotic event and not in a laboratory setting, so it was not possible to control any eventual disturbing factor without altering the inherent nature of the event. In order to have a consistent set of data for the age range (14–18 years old), only 79 pre- and 79 post-questionnaire (68 boys and 11 girls; $M_{age}$ = 15.36; $r$ = 1.49) were finally retained for the purposes of analysis.

### 1.3.3.3. *Materials*

The pre- and post-questionnaires were designed for the purposes of this study. All the questions were asked in a closed form so that participants' answers consisted of scores on a 5-point Likert scale. The questionnaire filled in by participants before the RoboParty® had four main sections (see Appendix 3). Section 1 consisted of four questions about the sample population (age, gender, team and school names) and was designed to collect students' profiles. Section 2 consisted of five questions designed to measure participants' familiarity with robots. For example: "How often have you watched or read something about robots?", "How often have you built or programmed a robot?". Section 3 consisted of one question: "How much does this picture make you think about robots?". This question was repeated for 10 items presented in the form of 10 different pictures, printed in color. The items presented in 10 pictures were selected because they were representative of the two ontological categories: living (i.e. *plants, animals and human beings*) and non-living entities (i.e. *single-function and multifunction machines, scientific instruments, toys*), which included some ambiguous entities (i.e. *anthropomorphic objects, animated characters, bionic components*). For each item, participants had to indicate on a 5-point Likert scale if the item made them think about a robot (1: "not at all" and 5: "a lot"). This section was designed to investigate whether and to what extent robots belong to living or non-living categories of entities according to students. Section 4 consisted of three questions designed to investigate the educational role(s) envisaged for the robot by participants (*an object to be constructed and programmed, a learning tool, a classmate*). Just as for the previous question, participants were invited to answer using a 5-point Likert scale (1: "not at all" and 5: "a lot").

The post-questionnaire (see Appendix 3) had the same format as the pre-questionnaire but it differed in the following respects: it did not include section 2 on familiarity with robots; furthermore, the pictures used in section 3 were replaced by different pictures, although they still represented the same categories of entities in order to prevent any potential habituation effect. The pre- and post-questionnaire had been tested previously by 30 volunteers at the Cité des Sciences et de l'Industrie, Paris. This first test aimed to assess the understandability of the instructions, the accessibility of the questions and how the proposed pictures (section 3) were interpreted, that is if the pictures selected for the pre- and post-questionnaire were representative of the categories we intended to investigate. 20 of the selected images were considered to be more than 80% "good examples" of our 10 categories. Subsequently, the questionnaires were refined and translated into Portuguese and English, the two official languages of the RoboParty®. In order to avoid biased responses to both the pre- and post-questionnaire, both the order of the items in each section and the order of the sections were randomized across the questionnaires.

#### 1.3.3.4. *Procedure*

Each participant who volunteered for the study was requested to fill in two questionnaires: one before the robot building and programming task and the other after the task. The pre-questionnaire was handed out in the form of a paper sheet to each team leader on the first RoboParty® day at the welcome desk. The team leaders were in charge of distributing these paper sheets, one for each member of the team. Participants wrote their answers directly onto the questionnaire sheet before opening the Bot'n Roll® kit, over a 15–20 minute period approximately. As soon as the participants had filled in the sheets, we collected them, with the help of volunteers from the University of Minho. Similarly, on the last day of the RoboParty® the team leaders were provided with the post-questionnaire, which was filled in individually by participants after they had built and programmed the robotics kit, over a 10–15 minute period during the last phase of the competition.

#### 1.3.3.5. *Data collection and analysis*

Responses to the pre- and post-questionnaire were used to create quantitative measures. Two types of raw scores were computed and used as a basis for compound scores:

1) *absolute scores*: scores given by the participants to each question;

2) *graded scores*: to assess the nuance of participants' representations, each participant's answer was categorized either as a graded answer, if the score provided was 2, 3 or 4, or as an absolute answer, if the score provided was 1 or 5.

Three compound scores were computed and used as dependent variables to evaluate participants' representations of robots before and after robot-making in an educational learning context and whether previous exposure to robots had an impact on these representations.

*Familiarity*: A global familiarity score was attributed to each participant. This score was the average of the scores obtained for the four familiarity questions in section 2 of the questionnaire ("How often have you watched or read something about robots?"; "How often do you play with robots?"; "How often have you built or programmed a robot?"; and "How many times in the past have you attended robot-related lessons, exhibitions, competitions, or other events?"). We considered that a score between 1 and 3 (3 not included) indicated a low level of familiarity with robots and that a score between 3 and 5 (3 included) indicated a medium–high level of familiarity with robots. According to the familiarity score obtained, two independent groups were created.

*Essentialist versus non-essentialist robot categorization*: We attributed a graded score to each participant consisting of the number of graded answers to the question "How much does this picture make you think about a robot?" (section 3 of both questionnaires). Similarly, a graded score was attributed to each question, calculated as the total number of participants who provided graded answers to each question. A higher graded score reflected a non-essentialist categorization, that is a more nuanced judgment about robots' membership in relation to the categories of living and non-living entities; in contrast, a lower graded score reflected an essentialist categorization, that is a more defined judgment about this.

*Shift among the educational roles envisaged for robots*: As was the case for robot categorization, a graded score was attributed to each participant, calculated as the number of the graded answers given by each participant to the question "In the classroom, would you prefer: (1) building and programming the robot, (2) using the robot as a tool to learn school subjects, (3) learning with the robot as a classmate?". Similarly, a graded score was attributed to each of the proposed three options, which was calculated as the total number of participants who provided a graded answer for each option. A higher graded score reflected a more nuanced attribution of a given educational role to robots. In contrast, a lower graded score reflected a more defined attribution of a given educational role.

Statistical analyses were performed with SPSS 21 with a significance level set at 0.05. An unpaired samples Student's $t$-test was carried out to compare familiar and unfamiliar groups with respect to robot categorization and educational role. A paired analysis of variance (ANOVA) with two variables (general linear model) – picture and moment (before and after robot-making); educational role and moment – was carriedout to compare the gradation and shift scores obtained at the pre- and post-questionnaire. A paired samples Student's $t$-test was conducted to assess whether any picture and any proposed educational role was given a different score in the pre- and post-questionnaire. A $\chi^2$ test was performed to compare the number of graded scores for each picture and each proposed educational role on the pre- and post-questionnaire, and Yule's $Q$ to examine the direction of the observed variation when significant differences were found.

## 1.4. Results

### 1.4.1. *Which representation of robots for familiar and unfamiliar students?*

The mean familiarity score was 2.49 ($r$ = 1.58). 44 (55.7%) participants obtained a familiarity score between 1 and 3 (3 not included), that is they demonstrated a low level of familiarity with robots. 35 (44.3%) participants obtained a score between 3 and 5 ($M = 4.11$; $r = 0.79$), which reflected a high level of familiarity with robots.

### 1.4.2. *The living and non-living items most frequently associated with robots (pre-questionnaire)*

For all participants, two items were most frequently associated with robots: bionic components and multifunction machines (Figure 1.1 and Table 1.1). These results were observed both for participants who were already familiar with robots and for participants who were not. However, familiar participants seemed to preferentially associate multifunction machines, rather than bionic components, with the robot (bionic components: $M = 4.06$; $r = 1.3$; multifunction machines: $M = 4.09$; $r = 0.85$). The opposite trend occurred for unfamiliar participants (bionic components: $M = 4.14$; $r = 1.17$; multifunction machines: $M = 3.73$; $r = 1.17$).

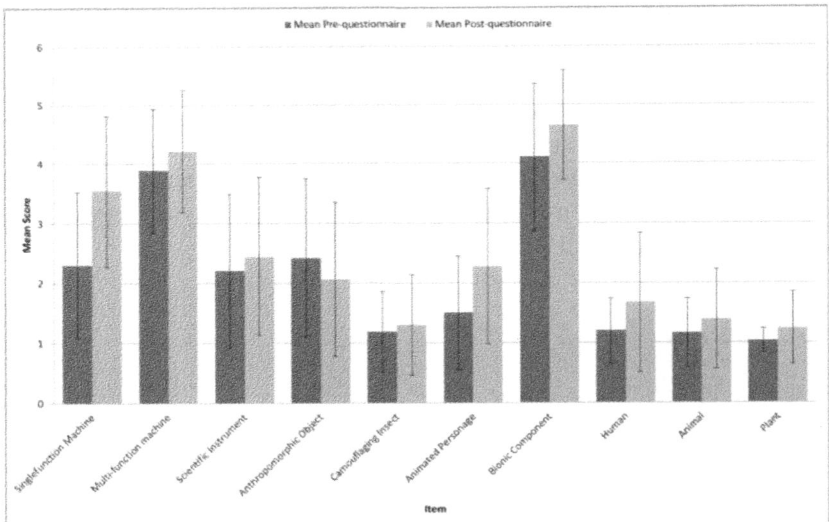


**Figure 1.1.** *Mean score for each picture on the pre- and post-questionnaire*

| *Item* | *Structural, functional and relational features* | *Category* |
|---|---|---|
| Single-function machine | A functional non-organic object, which is designed to serve one single purpose, and whose activation depends on users (e.g., a coffee machine) | Non–living |
| Multi-function machine | A functional non-organic object, designed to serve several purposes, whose activation depends on users (e.g., a computer) | Non–living |
| Scientific instrument | A functional non-organic object, designed as a tool for scientists (e.g., a microscope) | Non–living instrumental |
| Anthropo-morphic object | A social object with a human appearance, designed for playful interaction and on which it is possible to project human attitudes (e.g., a doll) | Non-living anthropomor-phic |
| Animated character | An object acting as a living entity, belonging to imaginary settings, designed to be watched rather than to interact with, and which is susceptible to personification (e.g., Pinocchio) | Non-living imaginary |
| Bionic component | An object designed to replace a living component, which is not organic and which becomes functional only when integrated with the living organism (e.g., a bionic eye) | Non-living ambiguous |
| Camouflage insect | An animal camouflaging with its surrounding environment, which is organic, autonomous, and which can be mistaken for stones, leaves, engendering unpredictable interactions etc. (e.g., a stick insect) | Living ambiguous |
| Plant | Vegetation growing with or without human care, which is organic, autonomous, but not responsive to human interaction (e.g., a tree) | Living vegetal |
| Animal | An animal growing with or without human care, which is organic, autonomous, and responsive to human interaction (e.g., a dog) | Living animal |
| Human being | A human individual, organic, requiring human care, becoming progressively autonomous and responsive to human interaction (e.g., a child) | Living human |

**Table 1.1.** *Description and categorization of 10 items represented in the 10 pictures in section 3 for the pre- and post-questionnaires. Participants were requested to look at these pictures and to answer the question "How much do these pictures make you think about a robot?" by giving a score on a 5-point Likert scale*

### 1.4.3. *Gradation in robot categorization: essentialist versus non-essentialist stance (pre-questionnaire)*

The mean graded score for all participants for the pre-questionnaire was 3.43 ($r$ = 1.52). This low score suggests an essentialist categorization of robots; the familiar group obtained a mean graded score of 3.51 ($r$ = 1.56) and the unfamiliar group obtained a mean graded score of 3.36 ($r$ = 1.50). No significant difference was found between familiar and unfamiliar groups using a Student's $t$-test ($t(79) = 0.181$; ns). Hence, familiarity seemed to have no effect on categorization: both familiar and unfamiliar participants seemed to opt for an essentialist categorization of robots.

### 1.4.4. *The educational roles most frequently envisaged for robots (pre-questionnaire)*

For all participants, the option "object to be constructed and programmed" appeared to be the most frequently envisaged educational role for robots, followed by "learning tool" and "classmate" (see Table 1.2; Figure 1.2). This result was observed for both familiar and unfamiliar participants.

### 1.4.5. *Gradation in the educational roles envisaged for a robot (pre-questionnaire)*

The graded score relating to the educational roles envisaged for robots for all participants was 1.41 ($r$ = 1.06). This graded score indicates a medium level of nuance in the given answers. The graded score for familiar participants was 1.37 ($r$ = 1.11), while the score for unfamiliar participants was 1.43 ($r$ = 1.02). No significant difference was observed between the two groups of participants on the pre-questionnaire ($t(79) = 0.61$; ns). Hence, neither the familiar nor the unfamiliar group seemed to have a definite preference for one educational role in particular.

| Role | All | Familiar | Unfamiliar |
|---|---|---|---|
| Object to be created | $M = 4.61$; $\sigma = 0.608$ | $M = 4.63$; $\sigma = 0.598$ | $M = 4.59$; $\sigma = 0.622$ |
| Learning tool | $M = 3.43$; $\sigma = 1.317$ | $M = 3.26$; $\sigma = 1.421$ | $M = 3.57$; $\sigma = 1.228$ |
| Classmate | $M = 2.75$; $\sigma = 1.531$ | $M = 2.80$; $\sigma = 1.549$ | $M = 2.70$; $\sigma = 1.534$ |

**Table 1.2.** *Means and standard deviations for scores given by familiar and unfamiliar participants to three educational roles of the robot envisaged in educational contexts*

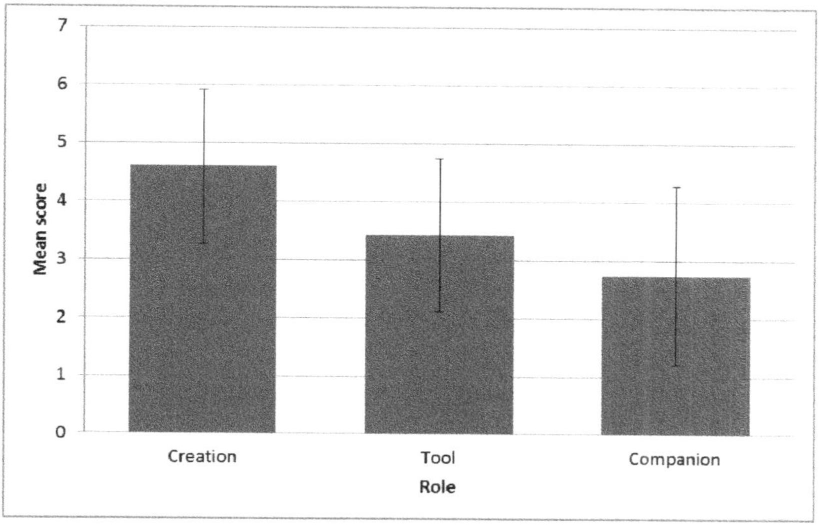


**Figure 1.2.** *Mean score for each educational role on the pre-questionnaire*

### 1.4.6. *The impact of building and programming a robot on students' judgment about the ontological status of robots*

In order to assess whether robot making had an effect on participants' categorization of the robot in their common-sense ontology, we initially performed a paired ANOVA with two variables (general linear model): picture (1–10) and moment (before and after robot making). The results showed a significant effect for picture ($F = 148.8$; $P < 0.001$) and moment ($F = 36.7$; $P < 0.001$).

A significant interaction effect ($F = 7.53$; $P < 0.001$) between picture and moment was observed, indicating that robot making had an effect on the representation of robots, when measured with our questionnaire.

In order to assess whether any picture was given a different score in the pre- and post-questionnaires we performed a paired Student's *t*-test on the scores given for the whole set of 10 pictures in both the questionnaires as post hoc comparisons. Table 1.3 shows the mean scores given by participants before and after building and programming robots, when they were asked to answer the question "How much does this picture make you think about a robot?" using a 5-point scale for each of the 10 displayed pictures. The results showed significant differences in the scores given for 7 of 10 of the presented pictures (Table 1.3).

More precisely, the mean score for the post-questionnaire increased for seven items: single-function machine, animated character, human being, bionic component, multifunction machine, plant and animal. In contrast, it remained relatively stable for the anthropomorphic object, scientific instrument and camouflage insect items (see Table 1.3 and Figure 1.3).

| *Item* | Pre-test mean SD ($N$ = 79) | Post-test mean SD ($N$ = 79) | $T$-test |
|---|---|---|---|
| Bionic component | $M = 4.10$ $\sigma = 1.236$ | $M = 4.65$ $\sigma = 0.934$ | $t(79) = 3.36$ $P < 0.001$ |
| Multi-function machine | $M = 3.89$ $\sigma = 1.050$ | $M = 4.22$ $\sigma = 1.034$ | $t(79) = 2.79$ $P < 0.01$ |
| Single-function machine | $M = 2.30$ $\sigma = 1.213$ | $M = 3.54$ $\sigma = 1.269$ | $t(79) = 7.94$ $P < 0.001$ |
| Scientific instrument | $M = 2.20$ $\sigma = 1.275$ | $M = 2.44$ $\sigma = 1.318$ | $t(79) = 1.28$ ns |
| Anthropomorphic object | $M = 2.42$ $\sigma = 1.317$ | $M = 2.06$ $\sigma = 1.284$ | $t(79) = -1.819$ ns |
| Animated character | $M = 1.51$ $\sigma = 0.946$ | $M = 2.28$ $\sigma = 1.290$ | $t(79) = 4.71$ $P < 0.001$ |
| Camouflage insect | $M = 1.19$ $\sigma = 0.662$ | $M = 1.30$ $\sigma = 0.837$ | $t(79) = 0.92$ ns |
| Human being | $M = 1.20$ $\sigma = 0.540$ | $M = 1.67$ $\sigma = 1.163$ | $t(79) = 3.65$ $P < 0.001$ |
| Animal | $M = 1.16$ $\sigma = 0.565$ | $M = 1.39$ $\sigma = 0.823$ | $t(79) = 2.58$ $P < 0.05$ |
| Plant | $M = 1.04$ $\sigma = 0.192$ | $M = 1.24$ $\sigma = 0.604$ | $t(79) = 2.69$ $P < 0.01$ |

**Table 1.3.** *Mean score given by participants (N = 79) for each picture (N = 10) in the pre- and post-questionnaire for the question "How much does this picture make you think about a robot?" (1–5 Likert scale) and comparison results (t-test). The items are ordered from the one most associated with a robot to the one least associated with a robot*

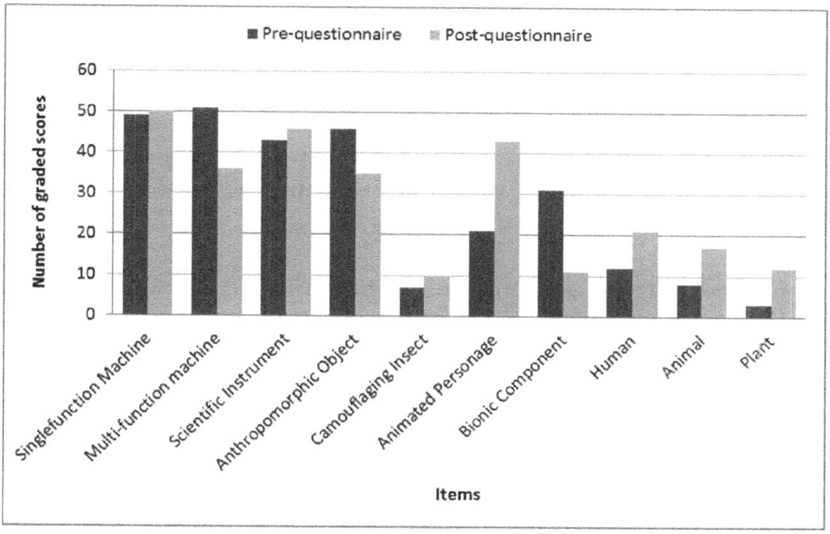


**Figure 1.3.** *Number of graded scores given for each picture on the pre- and post-questionnaires*

However, as in the pre-questionnaire, the bionic component ($M$ = 4.65; $r$ = 0.93) and multifunction machine ($M$ = 4.22; $r$ = 1.03) were the items that were most frequently associated with robots.

### 1.4.7. *The impact of robot making on graded versus all-or-nothing categorization*

In order to assess if building and programming robots had an effect on students' robot categorization, we compared the mean graded scores across the pre- and post-questionnaires. The mean graded score was 3.43 ($r$ = 1.52) in the pre-questionnaire and 3.56 ($r$ = 2.4) in the post-questionnaire, thus revealing that participants provided slightly more nuanced answers after making a robot.

Differences in the graded score obtained from the association between each picture and the idea of a robot that each participant had in their mind ("How much do these pictures make you think about a robot?"), before and after making a robot, may be considered as an indicator of how essentialist versus non-essentialist their assignment of the robot to living and non-living – and potentially to a new third category of entities – was. The paired Student's *t*-test performed on the graded scores for the whole set of pictures showed no significant difference across the pre- and the post-questionnaires ($t(79) = 0.043$; ns). Overall, participants' answers did not appear to become significantly more graded after making the robot.

Furthermore, we sought to investigate whether this result held for each picture individually. To this end, we performed a $\chi^2$ test in order to compare the number of graded scores for each picture on the pre- and post-test questionnaires. In accordance with the limits of the $\chi^2$ test, we also performed a *t*-test between the mean number of graded scores on the pre- and post-questionnaires when there were fewer than five graded scores on the pre- or post-questionnaires. The results showed that, although the average graded score for the 10 pictures taken all together remained globally constant between the pre- and post-questionnaires, the distribution of these scores varied across the pictures (Table 1.4 and Figure 1.3).

Specifically, we observed that for 7 of the 10 items, creating a robot from scratch had an effect on the number of graded scores given; this was the case for the multifunction machine, anthropomorphic object, animated character, bionic component, human being, animal, and plant (see Table 1.4). In contrast, no significant differences in the number of graded scores were observed before and after the robot making task for the remaining three items: single-function machine, scientific instrument, and camouflage animal (Table 1.4).

Finally, in order to assess whether the observed association could be interpreted in terms of an increase or a decrease in the number of graded categorizations in the post-questionnaires, we performed a Yule's *Q* test on the seven items, which were assigned significantly different graded scores in the pre- and post-questionnaires (Table 1.4). The results indicate that participants assigned a lower graded score for bionic component, multifunction machine and anthropomorphic object, and they assigned a higher graded score for the other items: animated character, animal, plant, and human (Table 1.4). In particular, the animated character and the bionic component appear to be the items which were most affected by the robot-making task, but affected in a different way: participants demonstrated a more graded and thus a non-essentialist categorization of the animated character; in contrast, they moved to a less graded and thus an essentialist categorization of the bionic component.

| Item | *N* of graded scores in pre-test | *N* of graded scores in post-test | $\chi^{2a}$ | Yule's *Q* |
|---|---|---|---|---|
| Single-function machine | $N = 49$ | $N = 50$ | $\chi^2 (1) = 0.054$ ns | – |
| Multi-function machine | $N = 51$ | $N = 36$ | **$\chi^2(1) = 12.447$ $P < 0.001$** | –0.370 |
| Scientific instrument | $N = 43$ | $N = 46$ | $\chi^2(1) = 0.459$ ns | – |
| Anthropomorphic object | $N = 46$ | $N = 35$ | **$\chi^2(1) = 6.297$ $P = < 0.05$** | –0.273 |
| Camouflage insect | $N = 7$ | $N = 10$ | $\chi^2(1) = 1.411$ ns | – |
| Animated character | $N = 21$ | $N = 43$ | **$\chi^2(1) = 31.392$ $P < 0.001$** | 0.535 |
| Bionic component | $N = 31$ | $N = 11$ | **$\chi^2(1) = 21.237$ $P < 0.001$** | –0.599 |
| Human | $N = 12$ | $N = 21$ | **$\chi^2(1) = 7.959$ $P < 0.01$** | 0.338 |
| Animal | $N = 8$ | $N = 17$ | **$\chi^2 = 11.266$ $P < 0.001$** | 0.417 |
| Plant | $N = 3$ | $N = 12$ | **$M_{Pre} = 0.04$ ($\sigma = 0.192$) $M_{Post} = 0.15$ ($\sigma = 0.361$) $t(79) = -2.584$ $P < 0.05$** | – |

[a]: A T-test was performed if the number of graded scores was lower than 5 in the pre- or post-test questionnaires.

**Table 1.4.** *$\chi^2$ and Yule's Q results for the graded scores for 10 pictures on a 1–5 Likert scale in response to the question "How much do these pictures make you think about a robot?"*

In summary, after having assembled and programmed the robotics kit, participants demonstrated a more nuanced positioning of robots in relation to animated characters, but a more clearly defined positioning of robots in relation to bionic components (Table 1.4).

### 1.4.8. *Does familiarity with robots influence their categorization?*

In order to assess whether participants who were unfamiliar with robots would change their categorization of robots after robot making more drastically than those who were familiar, we performed a paired Student's *t*-test on the mean graded scores between pre- and post-questionnaires for both the unfamiliar and familiar groups.

In the pre-questionnaire, the mean graded score for the familiar group was 3.51 ($r$ = 1.56), while for the unfamiliar group it was 3.36 ($r$ = 1.50). In the pre-questionnaire, the score was 3.57 ($r$ = 2.82) for the familiar group and 3.55 ($r$ = 2.12) for the unfamiliar one. These scores thus reflect a slightly higher level of gradation in participants' answers. A paired Student's *t*-test did not reveal any significant difference in graded scores between pre- and post-questionnaire, both for the unfamiliar ($t(44)$ = –0.579; ns) and familiar groups ($t(35)$ = –0.106; ns). The results of the $\chi^2$ test, comparing the number of graded scores for each item before and after robot making, indicate that the scores given by unfamiliar participants tended to stay stable across the pre- and post-questionnaires, except for half of the items. Similarly, responses by participants who were already familiar with robots tended to stay stable across pre- and post-questionnaire, except for three items (Table 1.5 and Figure 1.4.).

Unfamiliar participants have a higher number of graded scores to the animated character and plant and fewer graded scores to the multifunction machine, scientific instrument and bionic component. In contrast, participants who had previous exposure to robots gave a higher number of graded scores to the animated character and fewer graded scores to the multifunction machine and anthropomorphic object. One of the items, the bionic component, seemed to really differentiate the two groups of familiar/unfamiliar participants: although the number of graded scores was constant between the pre- and post-questionnaires for the familiar group, it tended to be markedly lower for the unfamiliar group in the post-questionnaire (Table 1.5 and Figure 1.4).

| Items | $\chi^{2a}$ | | Yule's Q | |
|---|---|---|---|---|
| | *Unfamiliar* | *Familiar* | *Unfamiliar* | *Familiar* |
| Single-function machine | $N_{Pre} = 24$<br>$N_{Post} = 28$<br>$\chi^2 = 1.467$; ns | $N_{Pre} = 25$<br>$N_{Post} = 22$<br>$\chi^2 = 1.260$; ns | – | |
| Multi-function machine | $N_{Pre} = 29$<br>$N_{Post} = 21$<br>**$\chi^2 = 6.474$; $P<0.05$** | $N_{Pre} = 22$<br>$N_{Post} = 15$<br>**$\chi^2 = 5.997$; $P<0.05$** | –0.386 | –0.323 |
| Scientific instrument | $N_{Pre} = 20$<br>$N_{Post} = 26$<br>$\chi^2 = 3.300$; ns | $N_{Pre} = 23$<br>$N_{Post} = 20$<br>$\chi^2 = 1.141$; ns | | – |
| Anthropo-morphic object | $N_{Pre} = 24$<br>$N_{Post} = 21$<br>$\chi^2 = 0.825$; ns | $N_{Pre} = 22$<br>$N_{Post} = 14$<br>**$\chi^2 = 7.820$; $P<0.01$** | – | –0.435 |
| Camouflage insect | $M_{Pre} = 0.07$ ($\sigma = 0.255$)<br>$M_{Post} = 0.11$ ($\sigma = 0.321$)<br>$t(44) = 0.813$; ns | $M_{Pre} = 0.11$ ($\sigma = 0.323$)<br>$M_{Post} = 0.14$ ($\sigma = 0.355$)<br>$t(35) = 0.329$; ns | – | – |
| Animated character | $N_{Pre} = 13$<br>$N_{Post} = 25$<br>**$\chi^2 = 15.722$; $P<0.001$** | $N_{Pre} = 8$<br>$N_{Post} = 18$<br>**$\chi^2 = 16.204$; $P<0.001$** | 0.517 | 0.562 |
| Bionic component | $M_{Pre} = 0.43$ ($\sigma = 0.501$)<br>$M_{Post} = 0.07$ ($\sigma = 0.255$)<br>**$t(44) = 4.200$; $P<0.001$** | $N_{Pre} = 12$<br>$N_{Post} = 8$<br>$\chi^2 = 2.029$; ns | – | – |
| Human being | $N_{Pre} = 8$<br>$N_{Post} = 12$<br>$\chi^2 = 2.444$; ns | $M_{Pre} = 0.11$ ($\sigma = 0.323$)<br>$M_{Post} = 0.26$ ($\sigma = 0.321$)<br>$t(35) = 1.406$; ns | – | – |
| Animal | $N_{Pre} = 6$<br>$N_{Post} = 10$<br>$\chi^2 = 3.088$; ns | $M_{Pre} = 0.06$ ($\sigma = 0.236$)<br>$M_{Post} = 0.20$ ($\sigma = 0.443$)<br>$t(35) = 1.966$; ns | – | – |
| Plant | $M_{Pre} = 0.05$ ($\sigma = 0.211$)<br>$M_{Post} = 0.11$ ($\sigma = 0.321$)<br>$t(44) = -1.354$; ns | $M_{Pre} = 0.03$ ($\sigma = 0.169$)<br>$M_{Post} = 0.20$ ($\sigma = 0.406$)<br>**$t(35) = 2.240$; $P<0.05$** | – | – |

[a]: A $t$-test was performed if the number of graded or absolute scores was lower than 5 in the pre- or post-test.

**Table 1.5.** *$\chi^2$ and Yule's Q results for graded scores on a 1–5 Likert scale given as an answer to the question "How much do these pictures make you think about a robot?" for 10 pictures*

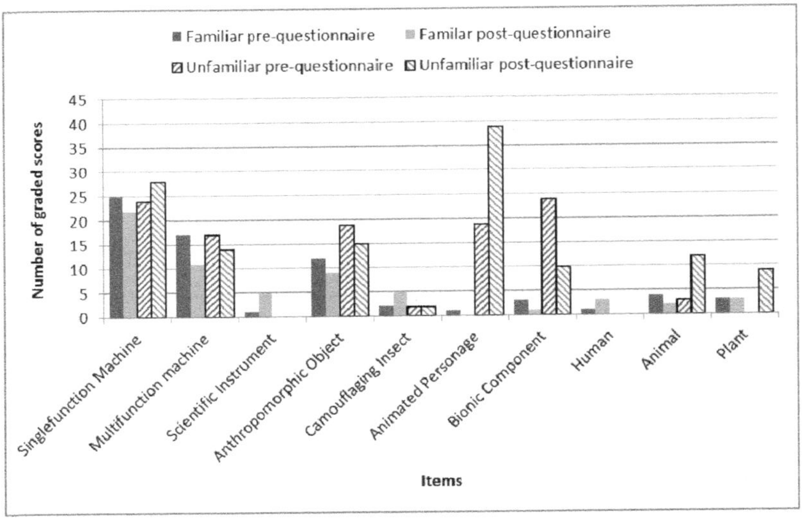


**Figure 1.4.** *Number of graded scores given for each picture on the pre- and post-questionnaire by familiar and unfamiliar participants*

### 1.4.9. *Dichotomous versus multiple categorization of robots*

We wished to assess whether the distribution of scores suggested (1) two groups of pictures in the pre-questionnaire – that is a clear living/non-living categorization of items in relation to the idea of a robot that participants had in mind – or (2) three groups of pictures in the post-questionnaire – that is a clear living/non-living/new entities categorization for the same items, again in relation to participants' mental images of robots.

Therefore, we performed an affinity propagation clustering algorithm [FRE 07] on the correlation matrices of scores given by participants on the pre and the post-questionnaires for each of the 10 pictures. Results did not show the emergence of a third category of items.

Two clusters (Table 1.6) emerged from the data, both in the pre and the post-questionnaires, showing that participants tended to make a clear distinction between "pure" machines (*single-function machine, multifunction machine, scientific instrument and bionic component*) and living or life-like entities (*human, animal, plant, camouflage insect and animated character*), in both the pre- and post-questionnaires. However, it has to be noted that one of the items, the anthropomorphic object, moved from the non-living cluster to the living entities cluster after the robot-making task (Table 1.6).

| | Pre-questionnaire | | Post-questionnaire | |
|---|---|---|---|---|
| Clusters | *Cluster 1* | *Cluster 2* | *Cluster 1* | *Cluster 2* |
| Single-function machine | X | | X | |
| Multifunction machine | X | | X | |
| Scientific instrument | X | | X | |
| Bionic component | X | | X | |
| Anthropomorphic object | X | | | X |
| Camouflage insect | | X | | X |
| Animated character | | X | | X |
| Human | | X | | X |
| Animal | | X | | X |
| Plant | | X | | X |

**Table 1.6.** *Results of an affinity propagation clustering algorithm performed on the scores on a 1–5 Likert scale given as an answer the question "How much do these pictures make you think about a robot?" in the pre- and post-questionnaires*

### 1.4.10. *The impact of robot making on the educational roles envisaged for robots*

To assess whether robot making had an impact on a shift between different educational roles for robots, we performed a paired ANOVA on two variables (general linear model): role (object to be constructed and programmed, learning tool and classmate) and moment (before and after robot making). The results indicated no significant interaction effect between role and moment ($F$ = 1.22; ns). Therefore, participants did not appear to change their preference for the different educational roles for the robot after they had built and programmed it (Figure 1.5 and Table 1.7).

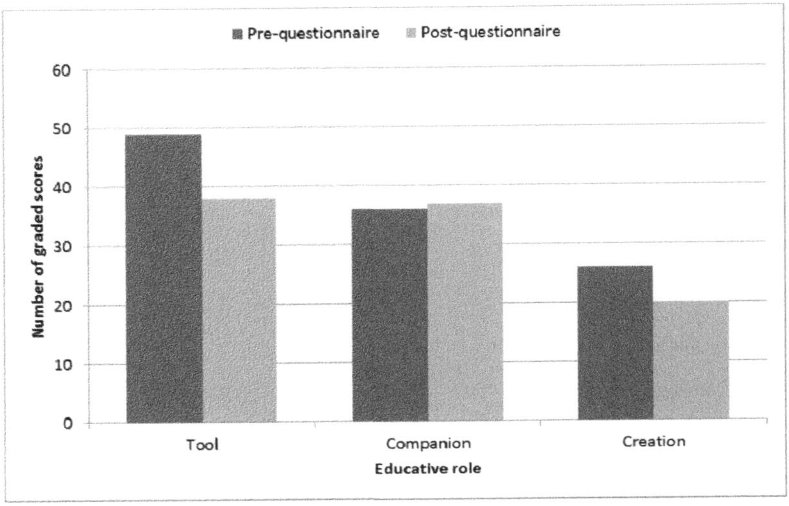


**Figure 1.5.** *Number of graded scores given for each educational role envisaged for robots on the pre- and post-questionnaires*

| **Educational role** | $\chi^2$ | **Yule's *Q*** |
|---|---|---|
| Tool | $N_{Pre} = 49$<br>$N_{Post} = 38$<br>$\chi^2 = 6.503$;<br>***P*<0.05** | –0.276 |
| Classmate | $N_{Pre} = 36$<br>$N_{Post} = 37$<br>$\chi^2 = 0.051$; ns | – |
| Creation | $N_{Pre} = 26$<br>$N_{Post} = 20$<br>$\chi^2 = 2.064$; ns | – |

**Table 1.7.** *$\chi^2$ and Yule's Q results for the score given in response to the question "Would you like your robot to be a) a tool, b) a classmate or c) an object to be created"*

### 1.4.11. *The impact of robot making on shift between the educational roles envisaged for robots*

In order to assess whether building and programming robots had an effect on the educational role participants attributed to the robot, we compared the mean graded scores for participants on the pre- and post-questionnaires. The mean graded score related to the educational roles envisaged for robots and could be scored from 0 to 3; the mean score was 1.41 ($r$ = 1.06) in the pre-questionnaire and 1.06 ($r$ = 1.07) in the post-questionnaires.

The paired Student's $t$-tests performed on the average scores for all the three roles revealed a significant difference between the graded scores obtained before and after the robotics activity ($t(79)$ = 2.2; $P < 0.05$). Robot making appeared to have an impact on the graded score given for the different educational roles envisaged for the robot (the robot as an object to be constructed and programmed, the robot as a tool for learning school subjects, the robot as a classmate).

In order to assess whether this result held for each of the three proposed educational roles separately, we performed a $\chi^2$ test to compare the number of graded scores given by participants for each of the three options in the pre- and post-questionnaires (see Figure 1.5). The results of the $\chi^2$ test indicated that robot making had a weak impact on participants giving graded scores in the post-test (Table 1.7): only the gradation of the score for the "learning tool" option significantly changed across the pre- and post-questionnaires.

Finally, in order to assess whether we observed a prevalence of graded scores in the post-questionnaires, we performed a Yule's $Q$ (Table 1.7) for the question for which scores differed significantly across the pre- and post-questionnaires, that is on the scores given for the "tool for learning school subjects" option. This analysis shows that participants gave fewer graded scores in the post-questionnaires for this option indicating that they did not attributed multiple educational roles to the robot.

### 1.4.12. *Does previous experience influence the educational role attributed to robots?*

In order to assess whether previous exposure to robots determines a lesser striking impact of robot making on students' attribution of educational roles to robots, we carried out several analyses. In the pre-questionnaire the mean graded score for the familiar group was 1.37 ($r$ = 1.11), whereas the mean score for the unfamiliar group was 1.43 ($r$ = 1.02). In the post-questionnaire, the mean graded score for the familiar group was 1.37 ($r$ = 1.17), while that of the unfamiliar group was 0.82 ($r$ = 0.922). No significant difference between the two groups was found using a Student's $t$-test. A significant result was obtained instead for the unfamiliar group ($t$ = 3.14; $P < 0.01$). After robot making, unfamiliar participants thus appeared to provide less nuanced answers concerning the educational roles for robots, while the answers given by familiar participants seemed to stay stable.

We then verified whether these results were valid for each role separately. To do this, we performed a $\chi^2$ test to compare the number of graded scores before and after robot making and a Yule's $Q$ to examine the direction of the observed variation when significant differences were found (Table 1.8 and Figure 1.6).

Participants who were familiar with robots tended to give more graded scores for the "classmate" option after robot making, but no effect was found for the other two options ("object to be constructed and programmed" and "learning tool"). Participants who were new to robotics appeared to give fewer graded scores to the "learning tool" and "object to be constructed and programmed" options, while the number of graded scores given to the "classmate" option tended to remain constant (Table 1.8 and Figure 1.6).

To summarize, for familiar participants, only the role of classmate was affected in the sense that a higher number of graded scores were given. For unfamiliar participants, there were no changes across the pre- and post-questionnaires in terms of the number of graded scores; on the contrary, these participants appeared to have a more clearly defined idea about whether robots could be a learning tool or an object to be constructed and programmed.

| Educational role | $\chi^2$ Unfamiliar | Familiar | Yule's $Q$ Unfamiliar | Familiar |
|---|---|---|---|---|
| Tool | $N_{Pre} = 28$<br>$N_{Post} = 17$<br>**$\chi^2 = 11.884$; $P<0.001$** | $N_{Pre} = 21$<br>$N_{Post} = 21$<br>$\chi^2 = 0$; ns | –0.471 | – |
| Classmate | $N_{Pre} = 20$<br>$N_{Post} = 15$<br>$\chi^2 = 2.292$; ns | $N_{Pre} = 16$<br>$N_{Post} = 22$<br>**$\chi^2 = 4.145$; $P<0.05$** | – | 0.335 |
| Creation | $N_{Pre} = 15$<br>$N_{Post} = 8$<br>**$\chi^2 = 4.956$; $P<0.05$** | $N_{Pre} = 11$<br>$N_{Post} = 12$<br>$\chi^2 = 0.133$; ns | –0.399 | – |

**Table 1.8.** *$\chi^2$ and Yule's Q results for scores given on a 1–5 Likert scale as an answer given by familiar and unfamiliar participants to the question "Would you like your robot to be (a) a tool for learning school subjects, (b) a classmate or (c) an object to be created"*

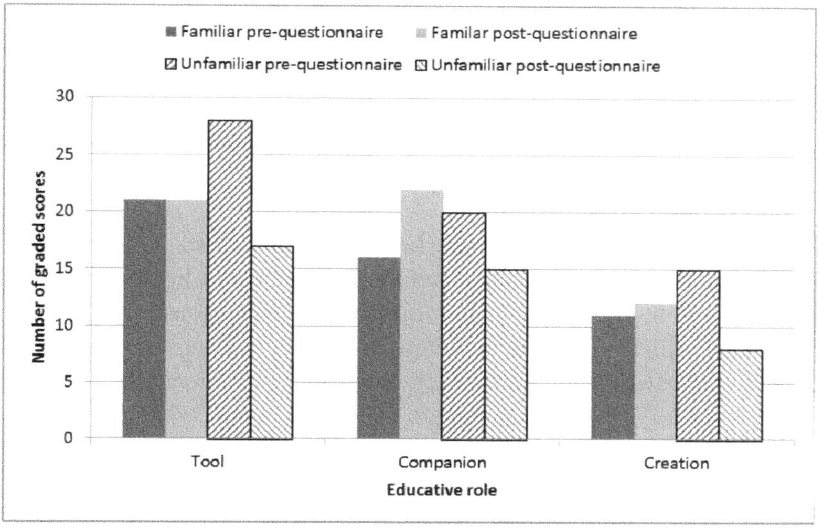


**Figure 1.6.** *Number of graded scores given for each educational role envisaged for robots in the pre- and post-questionnaires by familiar and unfamiliar participants*

## 1.5. Discussion

Prior studies have suggested that acquaintance with embodied artificial intelligences produces a more nuanced [SCA 95] and species-specific view of them [BER 08a]. In particular, the more young people become experienced with robots, the more they show a nuanced categorization of robots, as being somehow in between living and non-living entities [TUR 11]. This has led some authors to argue for the hypothesis that a new category is emerging, one that does not map onto humans, animals or artifacts (NOC hypothesis [KAH 12]). In this respect, ER kits possess an interesting peculiarity: in order for users to have a fully functional interaction with these robots, they have to build and to program them. Therefore, they continuously play the intermittent roles of creator and user, treating the robot as an animated and non-animated object at the same time [ACK 91, KAP 05, TUR 11].

These theoretical arguments and experimental findings have led us to question the status of robots for a sensitive target of users, that is students. The underlying idea was that, due to the particular status of robotics kits, the use of this kind of robot demands: (1) a non-obvious ontological categorization and (2) a continuous shift between different educational roles of the robot.

We thus carried out an experimental study where 79 participants completed two questionnaires, before and after getting involved in a 3-day robot-making event. The objective of these questionnaires was to assess which ontological category robots belong to and which educational role(s) they may cover, according to students.

To achieve this aim, we asked participants to associate their idea of a robot to a set of 10 images (representing living and non-living items) and to three kinds of educational roles (i.e. an object, a tool and a classmate). In particular, by using a system of graded scoring for participants' answers, we were interested to understand to what extent students' ontological and functional categorization of robots was essentialist (i.e. all or nothing) or non-essentialist (i.e. nuanced).

"Nuance" of categorization has been considered as an indicator of the sophistication of participants' representation of robots: the more the score was graded, the more the representation has been esteemed finely grained and species specific.

As expected, in the pre-questionnaire, participants mainly demonstrated an essentialist categorization of robots. This means that, before having concrete experience of constructing and programming a robot, students neatly assign robots to one specific ontological category. Additionally, results show that the more a student is familiar with robots, the more he/she tends to consider them as a sort of boosted computer; on the contrary, the less a student is familiar with robots, the more he/she tends to consider them as a hybrid item.

With regard to the educational roles envisaged for robots, the option "*object to be constructed and programmed*" appeared to be the most frequently attributed role for robots, followed by "learning tool" and "classmate". In line with our expectations, we observed no gradation in participants' scoring, that is no nuance in their judgments. This result can be explained by the fact that participants filled in the pre-questionnaire just before taking part in an event where they were going to construct and program a robot, so expectations about the event itself might have biased their answer. On the other side, contrary to our expectations and to the studies who have pointed out that familiarity with robots makes users assume different perspectives – once as engineer and once as psychologist [ACK 91] – and that they tend to attribute different roles to a robot [KAP 05, TUR 11]; no significant difference was noted between the familiar and unfamiliar groups.

To this concern, we have to acknowledge that even though familiarity with robots was calculated as a single indicator, familiarity should rather be considered as a composite indicator (e.g. having watched a robot in a science fiction movie does not engender the same kind of familiarity as having interacted with a real robot). In addition, we did not ask participants when they acquired such familiarity, whether it was at a recent time or longer ago. Thus, we cannot know how vivid the representations of robots they had in their minds actually were.

With concerns to the main focus of the study, which is the impact of robot making on the ontological and educational status of robots, results of the post-questionnaire show that robot making has an impact on categorization of robots among living and nonliving entities: as we expected and in line with the literature [SCA 95, BER 08], participants assigned a more graded score to four of the items represented in the pictures (*animated character, animal, plant and human*), whereas, contrary to our prediction, they assigned a less graded score to three of the items (*bionic component, multifunction machine and anthropomorphic object*), and their scoring stayed stable for the remaining items (*single-function machine, scientific instrument, camouflage insect*). Hence, it seems that when dealing with boundary or ambiguous items – that is non-living entities that borrow their appearance or functions from living entities (*the bionic component, anthropomorphic object and multifunction machine*) – participants made more clearly defined judgments about the degree of closeness between robots and these items after they had built a robot from scratch. In contrast, when dealing with living (*animal, plant, human*) or at least life-like items (the *animated character*), participants made more nuanced judgments about the degree of closeness between robots and such items. Furthermore, we noted that the bionic component and the multifunction machine were already the items most frequently associated with robots on the pre-questionnaire, with no particular gradation in their scoring. This result suggests that becoming acquainted with the functioning of a robot confirms previous beliefs about robots' place in common-sense ontology, but it also makes such beliefs more nuanced or sophisticated.

With regard to the differences between the familiar and unfamiliar, we observed that one of the items, the bionic component, seemed to mainly differentiate these two groups: while the number of graded scores stayed constant between the pre- and post-questionnaires for the familiar participants, there were much fewer graded scores for the unfamiliar participants. It thus seems that this kind of hybrid entity, like bionic components, becomes a sort of referential boundary when robots enter our ontology, a boundary which helps us to position robots among the range of the already known entities.

However, despite our predictions and previous studies (NOC [KAH 12]), no new specific category for robots seems to emerge from our data: participants seemed to make a clear distinction between "pure" machines (*single-function machine, multifunction machine, scientific instrument* and *bionic component*) and living or life-like entities (*human, animal, plant, camouflage insect* and *animated character*), both in the pre- and post-questionnaires. Interestingly, the anthropomorphic object was the only item that did not belong to the same cluster before and after the robot-making task: participants categorized it among the non-living entities on the pre-questionnaire and among the living ones on the post-questionnaire. This result suggests that while gaining experience in robotics does not lead to the creation of an NOC, we cannot exclude the fact that gaining familiarity with robots could progressively produce changes in our common-sense ontology, in terms of considering "alive" entities that we used to consider to "not alive". Of course, this point would require deeper investigation, and we consider it to be a plausible starting point for future research about the emergence of an NOC.

Finally, concerning the educational roles attributed by participants after the robot-making event, contrary to our predictions, there were no changes in nuance of judgment for unfamiliar participants; on the contrary, these participants appeared to have a more clearly defined idea about whether robots could be a learning tool or an object to be constructed and programmed. On the other hand, consistent with our prediction and to current literature on the issue [KAP 05, TUR 11], the role of classmate received a higher number of graded scores from familiar participants. This result can have two possible explanations. First, we shall remark that the robot the participants dealt with during the RoboParty® was a robotic kit; it thus may have been difficult to attribute to this kind of robot, created from a kit, a different role like that one of a classmate – at least, more difficult than if it had been another kind of robot, such as a humanoid, for instance. Second, investigations about people's ability to shift among different robot's roles can be ascribed to a more comprehensive debate concerning cognitive flexibility. Shifts dealing with the roles ascribed to a technological device are in fact made possible thanks to the flexibility of the human cognitive system, that is the ability to have multiple representations of the same object [CHE 06].

Familiarity and expertise are often said to affect cognitive flexibility, and thus the resulting ability to perform perspective shifts [SAN 13]. According to a number of authors (for an exhaustive review on this subject, see [CAÑ 05]) and contrary to our hypothesis, familiarity and expertise could cause cognitive inflexibility, since an individual who is familiar with or expert about something changes his/her perspective less often than a novice. Our experiment seems to confirm these findings by showing that once we acquire a comfortable level of expertise in relation to the role of the robot we tend to stick to it, rather to envisage another possible role.

Overall, our results show that, on the one hand, the more students get acquainted to robots, the more they tend to a nuanced categorization of robots in their common sense ontology (i.e. their assignment to an ontological category is not a matter of all or nothing, but rather a matter of degree). On the other hand, the more students get acquainted to robots, the more they tend to attribute a definite role to a robot. In this sense, as previous studies have already pointed out [SLO 03, KAP 05, TUR 11] robots share something with living entities and something else with non-living ones: (1) as living entities, they are susceptible to non-essentialist categorization; (2) as non-living entities, they need to have a precise role or function in order for people to interact with them. Moreover, while no third ontological category, beyond those of living and non-living entities, seems to emerge in the students' representation, one of the items presented in the task-rating pictures has passed from the non-living to the living category in the post-questionnaire. This might be interpreted as a minor cue of a prospective redefinition of classical ontological categories.

## 1.6. Conclusions, limits and perspectives

Our investigation on the impact of robot making upon robot representation points out that acquaintance with robots may have an impact on the place assigned to robots by students in their common-sense ontology, while their point of view about the roles served by the robot stay mostly unchanged. Hence, when providing judgments about robots, while ontological status seems to admit degrees, educational status seems not to admit degrees.

We consider this an interesting clue on the way toward answering the main question of this chapter, which is how firm robot's representations are. In fact, even if further studies are of course designed for a deeper understanding of the issue, we cannot exclude the possibility that the early exposure of students to robotic technologies, accompanied by a massive growth in the use of technology within daily tasks, in private and public environments, could lay the basis for a culture in which our common-sense ontology is slowly redesigned, until we reach an "environmental generational amnesia" [KAH 09]: a sort of illness of the new generation, who will forget preexistent natural and artificial categories and continuously elaborate new boundary categories.

On the contrary, when it comes to a robot's role, we might say that, in order for the user–robot interaction to be meaningful and efficient, the robot should have a precise function or role.

As frequently happens with studies investigating new research topics, our study had a number of different limitations in terms of materials and procedure. First, the 10 pictures displayed in both the pre- and post-questionnaires presented slight differences in form and content that could have influenced participants' answers. In terms of formal differences, some pictures were drawings while others were photos. Photos and drawings may trigger distinct interpretations of the items, which can, for example, be understood as real (photos) and imaginary (drawings) entities. However, because we were aware of such bias in the visual supports, we took care to maintain a one-to-one correspondence between the pre- and post-questionnaires: if an item (e.g. *an animal*) was presented in a photo in the pre-questionnaire (e.g. a photo of a dog), it was presented in a photo (e.g. a photo of a horse) in the post-questionnaire as well; in the same way, if an item (e.g. a single-function machine) was presented as a drawing (e.g. an illustration of a computer) in the pre-questionnaire, it was depicted using a drawing (e.g. a comic strip depicting a PC) in the post-questionnaire as well. In our future studies, this limitation could be overcome by displaying either only photos or drawings to represent all 10 items. Another interesting option that would provide a more rigorous semantics for the pictures has been proposed by Rakinson and Poulin-Dubois [RAK 01], whereby the category judgment concerns living and non-living entities with a high level of similarity (e.g. bird/airplane, animal/car). Moreover, although we took care to test such images in a previous test, in order to assess whether they were actually representative of living and non-living categories, we did not verify which ideas about living and non-living entities the participants of RoboParty® implicitly held, as Bernstein *et al.* [BER 08a] did in their study.

However, the emergence of dichotomous scoring for pictures displaying living entities versus pictures displaying non-living entities suggests the choice of items could be considered as valid. Second, the use of a Likert scale, which was deliberately selected because it is rather difficult to obtain yes/no answers to questions about robots [KAH 12], may have biased participants to express a graded judgment. This limitation, already highlighted by Kalish [KAL 95] in his study on categorization, is relevant in relation to our second hypothesis, concerning the increased number of graded scores in the post-test as one of the effects of robot making. As a possible solution, implicit methods could be envisaged (e.g. recording reaction time when participants are asked to associate their idea of robots with pictures representing a range of living and non-living category items).

Third, even though we referred to familiarity with robots as a single factor in the context of our study, this is, as discussed above (section 1.5), a resultant of several sources of acquaintance with robots, ranging from observation of robots (movies, cartoons, advertisement, etc.) to active interaction with robots (at school or at home). In future developments of this kind of study, we should therefore take into account different connotations of familiarity, and find more robust methods to assess whether such familiarity is a recent or older acquisition. This is even more relevant if we consider that, as pointed out by Dautenhahn [DAU 14, Chapter 38.2) differently to living species, the robots that populate our classrooms or robot competition today do not share a common evolutionary history, they are just very different robotic "species": "Thus, what we mean by 'robot' today will be very different from what we mean by 'robot' in a hundred years' time. The concept of a robot is a moving target, and we constantly reinvent what we consider to be 'robot'".

# 2

# Learning with Robotics: Functional and Social Acceptance of Robots

## 2.1. Functional and social acceptance of robots

Learning with robots demands acceptance of robots in order to share several functional and social tasks with them [FRI 14]. Acceptance is a sensitive topic of investigation (e.g. [KAP 04, RIE 08, HEE 09, WEI 09, YOU 09, SAL 10, WEL 10, DE 13]) in the field of human–robot interaction (HRI) research that addresses the design, understanding and evaluation of robotic systems, which involve humans and robots interacting through communication [MUR 10]. Traditionally defined as "the demonstrable willingness within a user group to employ information technology for the tasks it is designed to support" within the field of the User Acceptance of Information Technology [DIL 01], acceptance takes on a new significance when referred to robots.

Through an attentive survey of the literature, we can identify six connotations of robot acceptance: representational, physical, behavioral, functional, social and cultural[1] [GAU 16a]. Among these

1 *Representational acceptance* describes the influence of the mental models (stereotypes from science fiction, pre-existing mental models of technology, animals or children, etc.) on robot acceptance in terms of place attributed to the robot within common sense ontology and suitable roles of a robot. *Physical acceptance* defines the embodiment features of the robot – its morphology (anthropomorphic, zoomorphic, machine-like), size (human or animal size *vs.* household appliance or toy size, etc.) materials (e.g. organic *vs.* mechanic), etc. – which could impact robot acceptance in

dimensions, functional and social acceptances appear to play a key role for effective interaction in educational contexts [PIC 05, ZAA 08, WEI 09, FRI 14].

Functional acceptance refers to the level of perceived ease of use, usefulness [HEE 09, WEI 09], accuracy [BEE 11, SCH 13] and innovativeness [ROG 95, KAP 05] of the robot. Social acceptance covers a variety of issues such as social presence [HAM 06b, RIE 13], perceived sociability – often overlapping with social abilities, social intelligence [HEE 10a] and social influence [NOM 06] of robots.

Interestingly, while representational [GAU 15] and cultural [LI 10] acceptance continuously vary with relation to the users' progressive familiarization to robots [BAR 07b] and while we are likely to accept that a robot might have a rough physical aspect [TUR 11], or to excuse the robot for an inappropriate behavior as we would do with a child [YOU 09], we are much less likely to accept a useless robot [KAP 05] and we can be disappointed by a robot that lacks social abilities [SHA 09, HEE 10].

Today, robots endowed with functional and social abilities can be distinguished into two categories on the market [SHA 09], with distinct purposes and labels, e.g. "utilitarian" and "hedonistic" robots [LEE 11]. But the robotic industry is definitely moving toward robots that are both functional and social [SHA 09] and researchers have argued that user behavior will evolve accordingly: the more efficient a robot is, the more people would be likely to trust them on all levels, including a social level [YOU 09]. Indeed, contemporary research points out that, when it comes to real interaction situations, functional and social acceptances share a fundamental common ingredient:

---

terms of likeability and credibility of the robot. *Behavioral acceptance* concerns those behaviors – for example proxemics (i.e. intimate *vs.* public distance, direction of approach, etc.), and communication (e.g. verbal, gestural, etc.) – which can affect acceptance in terms of believability and fluency of interaction. Finally, *cultural acceptance* refers to fundamental issues of a given culture, for example appetite for technology, social care, educational values, etc. which might alter acceptance in terms of intention to use and duration of use.

*users' trust in the robot* [SHI 05, HAN 11]. Trust is thus increasingly employed to assess the quality of HRI [LEE 04, HOF 09, YAG 11, KAN 12, BRU 13, SCH 13, VAN 14] and, as it emerges from this literature, is considered a valid indicator of a robot's functional and social acceptance [YAG 11].

For example, suppose a student asks its robot to help in carrying out measurements for a science project. In this case, the student would have to trust the robot as a functional agent (functional acceptance). Now, suppose the same student asks the robot, which is an indispensable piece of equipment, to bring in a swimming pool bag (a bathing-cap? A pair of flip-flops?). In this case, the person would have to trust in the robot as a social agent (social acceptance). The question is will the person trust the robot's answer in both these cases? As shown in our example, functional and social acceptances are characterized by the same dynamics, but apply to very different issues, are based on different required knowledge such as technical estimation versus human care, and presuppose very different built-in knowledge and performance of the robot.

However, so far, the research models developed in the last decade to assess functional acceptance of robots [HEE 09, WEI 09, BEE 11, FRI 14] mainly take up their conceptual frame from older technology acceptance models [DAV 89, ROG 95, VEN 03], thus neglecting the specificity of the robot as a social agent. Several studies, however, have validated a significant set of indicators of social acceptance such as performance expectations, attitude toward technology and anxiety [WEI 09]; see [HEE 10a] for a review of those indicators.

But while those indicators can be useful to estimate the users' willingness to accept robots in an early phase of the interaction, trust can be a more eloquent indicator for daily and sustained interactions, in which users rely on robots to take decisions about functional and social issues. To explain our position, let us go back to the science project and swimming pool example.

This kind of daily activity (i.e. exchange of information, taking decisions, affirming one's point of view, accepting compromise) can be considered as problems [POP 91]. To solve these problems, we need to retrieve and apply knowledge that relates to either scientific rational thinking or natural thinking, the latter often being called common sense. These two kinds of thinking are based on different processes, finalities and content [PIA 67, GUI 99, JAC 13]. Scientific thinking seeks the truth through logical demonstration and it is supposed to result in a true solution [PIA 67, POP 91]. Here, the decision-making process relies on the examination of the objective knowledge on the physical world. The decision thus responds to the physical world itself and it is validated by the implementation of the rules (in our example, these rules are represented by conventional measures). On the other hand, natural thinking or common-sense (beliefs, opinions) works according to a subjective and/or social logic, whose finality is the achievement of a pragmatic solution, that is one of the more appropriate solutions in a given context [GUI 99]: here, the decision-making process relies on subjective knowledge and social norms [ABR 94a, ABR 94b]. The decision thus responds to personal evidence, which is connected to the common-sense knowledge acquired through everyday experience.

Now in order to accept the robot's help with relation to the first problem (the measurements for the science project), the student needs to rely on the robot's scientific knowledge and technical ability to provide an objective judgment on a functional question, relying on its technological equipment (e.g. its force sensors). On the contrary, in the second case (the choice between the bathing-cap and the flip-flops), the student needs to rely on the robot's common-sense knowledge. So, in both cases, there is an underlying inferential process by which the person bases his/her trust on the robot's knowledge or "savvy" [YOU 09] called into the decisional process.

Functional savvy refers thus to the robot's ability to efficiently operate by ensuring useful and accurate performances with relation to the functions it was designed for [WEI 09, HEE 10, FRI 14]. The tacit assumption about functional savvy is that, just like computers, the robot is equipped with scientific knowledge about the physical world and with powerful instruments (e.g. sensors) to make much more precise measurements than a human can. This ability is considered a necessary condition by users when deciding to employ a robot [KAP 05, HEE 09, BEE 11].

Social savvy describes the robot's ability to fit the social structures and activities of a given context [YOU 09] according to its role in the interaction [WEL 10]. The tacit assumption about social savvy is that the robot possesses common-sense knowledge, that is knowledge of situations, behavioral scripts and current norms that are in use [HAM 06a, DAU 07]. Therefore, within the limited scope of this research, we define functional acceptance as users' trust in the robot's functional savvy and social acceptance as users trust in robot's social savvy.

However, caution should be used: as mentioned, social savvy requires an adaptive context-dependent knowledge, which cannot be hardwired into the robot and demands complex learning algorithms [LOC 04]. Consequently, even if users do have expectations about a robot's social savvy [LOH 10, COE 12], at the present stage of robot development, these expectations are rarely confirmed [DUF 03, FON 03].

This explains why researchers are currently for the necessity of building measures that could assess not only trust in a calling robot's functional savvy but also fine psychological aspects of non-obvious trust in a robot's social savvy [HEE 10a].

Several subjective measures of trust in the HRI research field have been developed and are mostly based on self-reports (i.e. questionnaires). If these measures allow us to collect users' overt opinions, they also tend to induce a reflective mental posture, and are then limited in their capacity to register those spontaneous opinions and inner beliefs that could allow us to better understand on which robot knowledge the users base their trust and which are the possible relations that they establish between the functional and the social savvy.

An alternative and complementary approach to test trust in machines comes from the classical experimental paradigm of the media equation theory [NAS 95, NAS 96, NAS 00]. According to this theory, when engaged in collaborative tasks, people tend to unconsciously accept computers as social entities: they trust in answers provided by the computer and they conform to its answers. Then, Nass and his colleagues introduced the concept of conformation to investigate users' spontaneous or "mindless" response to computers as an objective measure to evaluate levels of trust in a machine and its acceptance as a potential partner. Likewise, one of the research challenge today is thus to design an experimental paradigm that enables to register those mindless reactions toward robots, which are susceptible of revealing users trust in robot, with particular attention for trust in its functional and/or social savvy.

For this reason, we have set the focus of our inquiry on the observation of individuals' trust behaviors as an indicator of functional and social acceptance. Furthermore, inspired by the work of Nass and his colleagues, we have proposed to adapt the well-known media equation theory paradigm[2] by employing users' conformation to robots' decisions as a new objective measure of human–robot trust during a human–humanoid decision-making task under uncertainty [YU 15].

By confronting users with tasks where they have to verbally express their decisions about functional and social issues, we assess whether the fact of experiencing perceptual uncertainty (e.g. evaluating the weight of two slightly different objects) and sociocognitive uncertainty (e.g. evaluating which is the most suitable item in a specific context) leads users to withdraw their own decision and conform to the robot's decision.

Moreover, relying on those studies where it was proven that trust can vary according to individual and contextual differences [HAN 11], we are also interested in identifying a set of factors that are likely to correlate with trust in a robot's functional and social savvy, such as individuals' propensity to control [BUR 79], a fear of being influenced [NOM 06] and the context (competitive vs. collaborative) in which the decision-making task involving the person and the robot takes place.

2 The media equation paradigm was transposed from human–computer interaction to human–robot interaction in a study by Kidd [KID 03]. However, our research intention is to adapt this paradigm to the specificity of the robot as a functional and social technology rather than to transpose it: instead of being confronted with a unique task such as the desert survival problem, our participants are confronted with two tasks, respectively, functional and social.

## 2.2. Trust as a fundamental indicator of acceptance

Trust can determine the overall acceptance of a system [PAR 97]. For this reason, there are a considerable number of studies seeking a better understanding of the human–robot trust dynamics (see [HAN 11] for a meta-analysis).

Broadly speaking, trust in system automation is classically defined as having confidence in the system to perform the appropriate action [BIR 04] with personal integrity and reliability [HEE 09]. Further definitions of trust took over from studies on human–human trust, commonly known as interpersonal trust [ROT 71, MAY 95], focusing on expectations, common goals, uncertainty and reliance as core elements of trust [BIL 12]. In this sense, trust describes the expectation that a robot will help achieve an individual's goals in a situation characterized by uncertainty and by reliance of the individual on the robot [LEE 04, HOF 09].

The empirical works on human–robot trust carried out in the latter half of the last decade (e.g. [LEE 04, BUR 07, YAG 11, DE 13]) employ a variety of measures that aim at detecting changes in users level of trust and also factors that might decrease or enhance it [SCH 13], as we will outline in the following chapters in order to argue the interest in considering conformation as an objective measure of trust in the robot's functional and social savvy.

### 2.2.1. *Commonly used measures of human–robot trust*

Two main categories of methods for assessing human–robot trust emerge in the literature: the first is based on subjective or explicit measures while the second one is based on objective or implicit measures.

Objective measures can be retrieved from quantitative behavioral data (e.g. response time) produced by individuals [HOF 05], whereas subjective measures (i.e. questionnaires, self-reports) can be retrieved from symbolic qualitative verbal data (e.g. opinions) produced by the individuals. If the former are developed in HRI in a limited way, the latter are widely used.

There are very few studies on the relationship between trust and behavioral cues. The most notable study in the field of HRI is from De Steno *et al.* [DES 12], which recorded face-to-face verbal interactions between human participants and a teleoperated Nexi robot, to identify sequences of non-verbal behaviors of the humans that could be used as an indication of trust toward the robot. They identified a set of four non-verbal cues (face touching, hand touching, arm crossing and leaning backward) that, performed in sequences, are indicative of untrustworthy behavior [DES 12, LEE 13].

Subjective measures, more frequently used in HRI research, can range from questionnaires including a few statements, e.g. "I would trust the robot if it gave me advice, I would follow the advice the robot gives me" [HEE 09], ranked by the users on a Likert scale, to more complex questionnaires addressing users' prior experiences, reputation of the robot, observed physical features and perceived functional capabilities of the robot [STE 06].

Furthermore, four types of scales have been developed to explicitly register attitudinal data on human–robot trust (see [SCH 13] for a detailed review). The *propensity to trust* is a scale conceived to measure a stable and unique trait of the individual and may provide useful insights to predict the initial level of trust in robots [YAG 11]. *Trustworthiness* is a scale related to the robot type, personality, intelligence, level of automation and perceived function [LEE 04, BRU 13] and can be used to measure the human–robot trust during their first approach. Similarly, the *affective trust* scale is more appropriate in the initial stage of the human–robot relationship: this scale refers to the individuals' attributions about the motives of a partner to behave in a given way [BUR 07]. But of course trust is also important to support a sustained interaction with robots. To this concern, the *cognition-based* trust scale is employed to observe the evolution of trust throughout time in terms of (1) understanding of robot functioning, (2) ability to interact with the robot and (3) expectancy of the robot [MER 08].

However, as we have already pointed out (see Introduction), unlike objective measures that are intended to assess immediate behavioral reactions, and which therefore may give access to people's inner beliefs, subjective measures are intended to assess verbally mediate reaction: the procedure itself of posing questions to users about their trust in the robot can eventually alter spontaneity and may not be revealing of the effective trust toward the robot. This is witnessed by the fact that objective measures often show low correlations with explicit measures [HOF 05].

Finally, to our knowledge, very few works include both subjective and objective measures to assess a human's trust of robots (e.g. [JOO 13]). In such works, the employed objective measures mostly consist of collecting information about the distance that the human maintains with respect to the robot during the interaction (i.e. proxemics). However, though proxemics measurements enable researchers to objectively register information about perceived safety and acceptance of the robot presence into the physical and social space [EDE 14], which are undoubtedly relevant for embodied aspects of functional and social acceptance, they have limited interest with regard to the specific research objectives of the present study, that focuses on trust in robot savvy, thus tackling more psychological aspects of functional and social acceptance such as human behavior in situations of uncertainty.

For those reasons, we have targeted conformation as a new type of measure to register human–robot trust during a decision-making task under uncertainty.

### 2.2.2. *Conformation as an innovative measure of human–robot trust*

The conformation experimental paradigm to investigate fine psychological aspects of people's trust in computer answers during a human–computer interaction comes from a set of studies which gave rise to the so-called media equation theory [NAS 94, NAS 95, NAS 96, NAS 00]. This theory argues that people unconsciously treat computers as social agents when asked to collaborate with them to achieve a decision-making task. As an example, in a study by Nass and Moon [NAS 00], adults confronted with a "Desert Survival Problem" [LAF 74] had to rank 12 items, in collaboration with a computer (e.g. a knife and a flashlight) in order of importance for survival in the desert. Unknown to the subjects, the computer's rankings were systematically dissimilar to each subject's ranking (for instance if a subject ranked an item as number 2, the computer would automatically rank that item as number 5 and so on). After having read the computer ranking, subjects were allowed to change their ranking or to leave it as it was. Results showed that participants who were told that the task achievement would depend on their collaboration with the computer and not on the human decision solely, or of the computer solely, trusted the quality of the information provided by the computer more, and consequently conformed more to the computer's answer.

In this sense, conformation, which means to withdraw one's decision in order to comply with the machine's decision, is a relevant measure to straightforwardly register whether the users trust in the agent's savvy more than in their own savvy.

Moreover, conformation as an objective measure has a specific relevance in HRI tasks where users are supposed to collaborate with, or delegate to, robots. These kinds of task typically require us to share duties and responsibilities, as well as to mutually adjust, so that it is important to know to what extent a user perceives the robot as trustworthy enough to let it partly take charge of such duties and to possibly correct human performance when needed.

### 2.2.3. *Factors influencing robot trust*

In their meta-analysis, Hancock and colleagues [HAN 11] examine the factors influencing robot trust and gather the results of this examination in a three-factor model. These three factors are as follows: human-related factors (e.g. personality traits), robot-related factors (e.g. beliefs about robots) and environmental-related factors (e.g. type of interaction in a given environment). Following this model, we have considered the following three factors that could specifically affect trust in functional and social savvy:

1) desire for control (DFC);

2) attitude toward social influence of robots;

3) type of interaction scenario.

#### 2.2.3.1. *Personality traits: DFC*

It has been proven that personality traits influence people's acceptance of technology in general [ALA 92] and robots in particular [WEI 08, LOO 10, FIS 11].

For example, extroverts tend to trust robots more than introverts [MCB 10]. Other personality traits such as "proactiveness", "social reluctance", "timidity" and "nervousness" have also been observed with relation to robot acceptance [WAL 05, MER 08, YAG 11]. For example, people who score higher on proactiveness keep a higher distance from the robot than others [WAL 05].

However, as the dynamics of trust itself implies that the "trustor expects the trustee to perform helpful actions irrespective of the ability to control him" [MAY 95], a crucial personality trait to examine in a study about human trust in robots is the DFC [BUR 79], that is the intrinsic and personal tendency of a person to control the events in one's life [BUR 79].

The relevance of this personality trait with respect to human–robot trust is even more evident when we consider that robotics engineering is progressing toward prototypes of robots, which are more and more autonomous and aspire to achieve a complete autonomy of robots to take care of functional tasks such as mowing the lawn or social tasks such as assisting elderly people [YAN 02, THU 04].

To this concern, several studies have preventively tested users' reactions to prospected scenarios where robots would entirely operate without human control. For example, it has been demonstrated that male participants do not let the robot come close if it is operating in autonomous mode [SYR 07, KAM 13, KOA 14]. Another study on approach initiation by Okita *et al.* [OKI 12] shows that participants feel reassured if the robot asks permission by verbal or non-verbal communication before starting an interaction. Finally, several works point out that users prefer a robot that, though being capable of adaptive behavior, still leaves the user in control [GIL 04, MAR 04, HEE 10].

These studies seem to bear witness that users may still not yet be totally open to full unconditional trust in a robot's functional and social savvy. However, to our knowledge, the role of individual DFC with relation to these low-trust behaviors has not been clarified in the current HRI literature. Nevertheless, it is reasonable to think that DFC can diminish users' willingness that robots be in charge of a task. With relation to our study, DFC might turn out to negatively correlate with participants' tendency to conform, that is "to give the robot the last word" in decision-making tasks.

#### 2.2.3.2. *Attitude toward robots: social influence*

Attitudes toward robots appear to be crucial for successful acceptance of a robot as well [WUL 14, DES 15]. By attitude, we mean any mental disposition matured through experience that might impact the reactions (behavioral, verbal, emotional) of the individual toward objects and situations [ALP 35, OST 69, REG 77]. While DFC constitutes a stable trait of personality, attitudes toward robot are more contingent: they can vary according to cultures [KAP 04, LI 10] as well as to the people's acquaintance to robots [BAR 07b] and they can change through time [SUN 10]. Then, observing the attitude makes it possible to predict the actual and potential trust behavior of an individual toward the robot.

In HRI research, attitudes toward robots are generally assessed through the use of tests and questionnaires. Among the most common attitude assessment tests, there is the *Negative Attitudes towards Robots Scale* (NARS [NOM 06]). This test was conceived to cover three types of negative attitudes: (1) toward situations of interaction with robots (e.g. "I would feel nervous operating a robot in front of other people"); (2) toward social influence of robots (e.g. "I am concerned that robots would be a bad influence on children") and

(3) toward emotions in interaction with robots (e.g. "I feel comforted being with robots that have emotions"). The NARS is thus composed of three different subscales, with each subscale including a list of statements that the respondent is invited to rate on a Likert scale from 1 (strongly disagree) to 7 (strongly agree).

Subscale 2 in particular focuses on negative attitudes toward the social influence of robots and it is thus especially relevant with relation to trust in social savvy. In this sense, negative attitudes could determine users' mistrust in the capability of a robot to fit social structures. Thus, we might expect that the more people show negative attitudes toward the social influence of the robots, the less they trust robots' decisions with regard to social issues. Furthermore, the NARS-S2 is also relevant to our methodological choice of employing conformation as a measure of trust, because it is reasonable to think that the more a person feels anxious about the idea of being influenced by a robot, the less he/she will tend to conform to its decision.

#### 2.2.3.3. *Type of HRI scenario*

Different proposals to reduce negative attitudes toward robots through simulation of real and imagined HRI scenario were put forward in recent studies. Kidd [KID 03], for example, simulated two types of real scenarios: one that demanded collaboration with robots and another in which participants learned pedagogical content by guidance from a robotic teacher. As an outcome, the collaboration scenario raised more trust than the teaching scenario: the users perceived the information provided by the robot of higher quality, and trusted the robot to a greater extent.

Kuchenbrandt and Eyssel [KUC 12] took over the experimental paradigm of "imagined contact" [CRI 12] from social psychology to assess whether the fact of asking people to close their eyes and imagine different types of HRI scenario has an effect on robot acceptance. Results showed that after having imagined an interaction with the robot, participants exhibited fewer negative attitudes and less anxiety toward robots. These effects were stronger for cooperative than for competitive and neutral imagined interaction. Conversely, in a subsequent study, Wullenkord and Eyssel [WUL 14] reproduced a similar scenario but obtained different results: participants who had imagined contact with a robot did not report more positive attitudes toward robots nor did they showed higher contact intentions with it than participants who had imagined contact with a human or with other kinds of technical devices.

It thus seems that giving users the possibility to project themselves in the context of the interaction has contrasting effects on the acceptance of robots.

However, simulating collaborative and cooperative scenario is especially suitable in relation to functional and social tasks where humans and robots are supposed to collaborate for joint objectives, and can thus be confronted with feelings of rejection and antagonism that can be predicted by analyzing their imagery.

## 2.3. Study 2: robot acceptance

### 2.3.1. *Aims and rationale*

On the basis of the examined literature concerning the dynamics of human–robot acceptance, trust as an indicator of acceptance and conformation as measure of trust (sections 2.2.1 and 2.2.2), the aims of our second experimental study are (1) to investigate whether participants conform their decisions to the robot's answer when experiencing uncertainty during a decision-making task with respect to functional and social issues; (2) to assess the speculative assumption according to which trust in social savvy requires trust in functional savvy; and (3) to investigate to what extent DFC, negative attitudes toward robot's social influence and a collaborative versus competitive scenario can be considered as factors influencing the robot's acceptance in terms of trust.

To this aim, we carried out a two-step experiment with 56 adult students and an iCub robot [IVA 13]. First, to gather information about their psychological robot-related profiles, participants were invited to fill in two questionnaires – DFC [BUR 79] and NARS [NOM 06] – 2 weeks before the day of interaction with the iCub robot.

Later, on the day of the interaction with iCub, to assess whether the trustors' mental image of the type of the interaction that they will have with the iCub impacts on their trust in the robot, the participants were assigned to one of the three groups and were asked to imagine a specific HRI scenario (i.e. collaborative, competitive and neutral) [KUC 12, WUL 14].

Finally, participants were confronted with two different decision-making tasks. In the first task (called functional task), participants were presented with a set of physical stimuli for which they had to compare a specific perceptual characteristic (i.e. the relative weight of two objects, the relative pitch of two sounds and a picture containing different colors). Immediately after, they were asked to decide which was the heaviest object, the most high-pitched sound and the predominant color. In a second moment, the experimenter presented iCub with the same set of physical stimuli and asked the robot the same question. Participants listened to the iCub answers, and were asked if they would like to keep their initial decision or change it. This first set of physical stimuli and related questions refer to the robot's functional savvy.

In the second task (called social task), participants were presented with set of paired items (e.g. a bathing-cap and a pair of flip-flops), and again asked to decide which of the items was the more suitable with relation to a specific social context (e.g. a swimming pool). In a second moment, the experimenter did the same with the iCub. After having listened to the iCub's answer, participants were allowed to change their decision. This second set of stimuli and related questions refer to the robot's social savvy.

Inspired by Nass and Moon's works on users' social responses to computers [NAS 95, NAS 96, NAS 00], we considered conformation, meant as participants' modification of their initial decision in compliance with robot answers, as a measure of trust.

We formulated the following five hypotheses:

HYPOTHESIS 2.1.– Participants trust the robot's functional savvy more than the robot's social savvy. As different studies pointed out, trust is a key component of HRI [HAN 11, SCH 13] and a valid indicator of acceptance [PAR 97, YAG 11]. Since Nass *et al.* [NAS 96] demonstrated that humans tend to express trust in computers through conformation to its answers, we expect that a similar phenomenon might appear during HRI, but according to the specificity of the robot as a functional and social technology. Such specificity implies that while users consider functional savvy as indispensible for accepting the robot [KAP 05, HEE 09, WEI 09, BEE 11, FRI 14], social savvy is rather a desirable feature to them [DAU 07, LOH 10, COE 12]. Thus, we predict that participants will conform more to functional than to social answers by the robot.

HYPOTHESIS 2.2.– Participants do not trust the social savvy uniquely: trust in the social savvy should be supported by trust in the functional savvy. As the literature shows, though participants expect robots to have social savvy, these expectations are often disappointed during real interaction [DUF 03, FON 03, COE 12]. Thus, we predict that no participant will conform to iCub answers in the social task uniquely.

HYPOTHESIS 2.3.– Participants who imagine a collaborative HRI scenario tend to trust the robot more than participants who imagine a competitive or neutral scenario. Previous studies [KUC 12] have demonstrated that imagined contact with robots makes people more confident toward a real robot, when the imagined scenario is a collaborative one. Consequently, we predict that participants having imagined a collaborative interaction with the robot will conform more to iCub answers both in the functional and social task.

HYPOTHESIS 2.4.– The more participants display a negative attitude toward the social influence of the robot, the less they trust the robot's social savvy. The statements of the S2 subscale of the NARS [NOM 06] mostly concern the negative feelings of people with relation to the possibility that the robot could influence or dominate them, and that they could depend on robots. Therefore, we predict that participants who score highly on the NARS-S2 will not conform to the iCub's answers in the social tasks, that is a high score on the NARS-S2 will be negatively correlated with the social conformation scores.

HYPOTHESIS 2.5.– The more participants show a strong DFC as a personality trait, the less they will trust the robot's functional and social savvy. Despite the fact that robotics technology is rapidly evolving toward fully autonomous artificial agents [YAN 02, THU 04], users still feel more confident in the interaction with robots if the robots are controlled by a human [GIL 04, MAR 04, SYR 07, HEE 10, OKI 12, KAM 13, KOA 14]. It is thus reasonable to think that individual differences in the DFC [BUR 79] might influence trust in robots. Therefore, we predict that the more participants score highly on the DFC test, the less they will conform to the iCub's answers both in the functional and social tasks, that is a high score negatively on the DFC will be negatively correlated with functional and social conformation score.

### 2.3.2. *Method*

#### 2.3.2.1. *Participants*

Fifty-six voluntary healthy adult students took part in the study: 37 women and 19 men. Nine were recruited from the Institute for Intelligent Systems and Robotics, 17 from the Institute for Distance Education of the Paris 8 University and 30 through a web "call for participants" placed on the Cognitive Sciences Information Network (RISC). They were all native French speakers, aged 19–65 (average age = 36.95; $\sigma$ = 14.32). As a token of appreciation, the participants received a gift voucher worth 10 euros. Participants signed an informed consent form to partake in the study and granted us the use of their recorded data and videos.

#### 2.3.2.2. *Experimental setting*

The experiments were conducted in the *Institut des Systèmes Intelligents et de Robotique* (ISIR, Paris, France) in the experimental room of the iCub robot. The experimental setup was organized as shown in Figure 2.1. The robot was standing on a fixed pole. A reflective wall (a plastic divider with reflective surface) was built to create a private space for them to interact with the robot, in particular to prevent the participants seeing the robot's operator. The participant and the experimenter were seated in front of the robot, with the experimenter being on the right of the participant. The position of the seats with respect to the robot was fixed and equal for all the participants. Between the robot and the experimenter, a LCD screen connected to a computer was used to display images related to the functional and social tasks. Two cameras were observing the participants: one camera was placed behind the robot on its left side, in such a way to observe the human face and upper body during the interaction with the robot; the other camera was placed laterally to take the scene as a whole, observing the overall behavior of the participants.

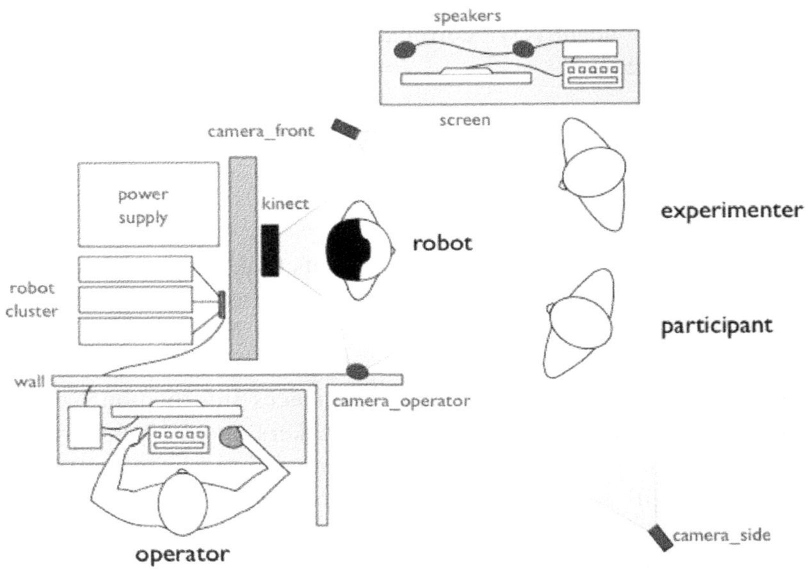


**Figure 2.1.** *Experimental setting. The human and the participant are seated in front of the iCub robot. An operator, hidden behind a wall and not visible by the participant, monitors the experiment and controls the robot to generate appropriate gestures and answers to questions*

*The iCub robot and the Wizard of Oz paradigm*: the participants were interacting with the iCub humanoid robot[3] [NAT 13]. The robot is approximately 104 cm high, weighs about 24 kg and has the shape of a 4 year old child. The robot was standing on a fixed pole so that it

3 The iCub humanoid robot is the outcome of the European RobotCub project: http://www. robotcub.org.

could not fall. As a safety measure, the robot was constantly monitored by the operator and controlled in impedance to make it compliant in case people touched it accidentally or intentionally during the experiment [FUM 12]. The experimenter was also able to stop the robot in case of urgency at any time using the robot safety button; however, there was no use of this measure during the experiments, as the interaction flowed without problems for all the participants.

Facial expressions and speech were enabled. During the experiments, the robot always assumed the same neutral/positive expressions to avoid confusing the participant or suggesting that the participant's actions could arouse a possible robot "emotional status".

*The robot was able to answer the questions of the functional and social tasks*: the verbal answers were preprogrammed in advance by the experimenter though the operator was able to type new sentences on-the-fly and make the robot speak in case of unforeseen questions by the participants.

This was made possible by implementing the Wizard of Oz paradigm (see [RIE 12] for a critical review). In the Wizard of Oz setting, the participants think they are interacting with an autonomous system, while in fact the system is partly or completely operated by an operator who is remotely in command of the robot. This paradigm allows the operator to control the robot's behavior in real time. To facilitate the control of the robot by the operator, we developed a graphical user interface (GUI) to quickly send high-level commands to the robot in a Wizard-of-Oz mode (WoZ) (Appendix 4).

#### 2.3.2.3. *Tasks and material*

The two online questionnaires and the tasks are detailed in the following.

*The questionnaires*: two online questionnaires were submitted to the participants.

The first was a French adaptation[4] of the NARS [NOM 06] (Appendix 4). Among the three subscales composing this test, we adopted the second subscale (negative attitude toward the social influence of robots) as a measure of attitude toward social influence of robots. This subscale includes five comments: (1) *I would feel uneasy if robots really had emotions;* (2) *Something bad might happen if robots developed into living beings;* (3) *I feel that if I depend on robots too much, something bad might happen;* (4) *I am concerned that robots would have a bad influence on children* and (5) *I feel that in the future society will be dominated by robots.* Subjects were required to answer on a Likert-type scale, from 1 (*strongly disagree*) to 7 (*strongly agree*).

The second questionnaire was a French adaptation of the DFC scale [BUR 79], which we used as a measure of participants' DFC. Twenty questions such as "*I'd rather run my own business and make my own mistakes than listen to someone else's orders*" composed this questionnaire (Appendix 4). Again, subjects were asked to answer on a Likert-type scale from 1 to 7.

The original questionnaire along with our French adaptation is reported in Appendix 4.

#### 2.3.2.3.1. The functional task

In the task designed to test the trust in the robot's functional savvy, the participant had to answer a series of questions about images (Figures 2.2 and 2.3), weights and sounds (Figures 2.4 and 2.5). The experimenter would ask each question to the participant, then to the robot and finally would ask the participant whether he/she would like to confirm or change his/her answer so that the participants and the robot could disagree or agree.

---

4 To our knowledge, although this test is widely used and validated in English speaking countries and in Asia (Japan, Korea, etc.), only one study employing NARS has been carried out in France [DIN 15a]. Validation by a group of 10 people was then implemented in order to ensure that the questions translated into French were appropriately understood.

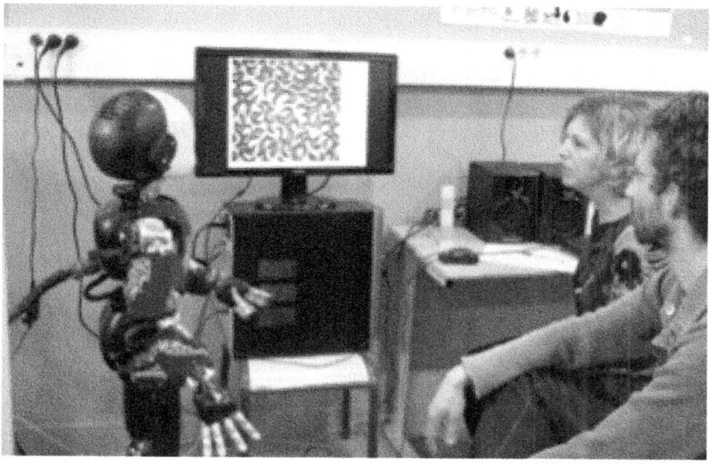

**Figure 2.2.** *Functional task and evaluation of images. The robot is gazing at the screen where the image to evaluate is shown*

**Figure 2.3.** *Functional task. The four images used for the evaluation of the dominant color*

**Figure 2.4.** *Functional task: the bottles used in the evaluation of weight: (1) two identical bottles of same weight, (2) two similar bottles of different colors and very different weight, (3) two similar bottles of almost the same weight, and (4) two different bottles of almost the same weight*

**Figure 2.5.** *Functional task: the participant evaluates the weight of the bottles (left); the experimenter gives the bottle to iCub for evaluating the weight (right)*

In the sound subtask, the experimenter asked "Which is the most high-pitched sound: the first or the second?". In the image subtask, the experimenter asked "Which is the dominant color in this image?". In the weight sub-task, the question was "Which is the heaviest object?".

Each subtask was composed of four evaluations, where a pair of items was compared. Among the four pairs of stimuli, three were ambiguous (i.e. *the two items slightly different*) and one was not ambiguous (*the two items were very different*). Ambiguous stimuli were introduced to assess users' behavior in situations of strong uncertainty (three subtasks are detailed in Appendix 4).

The strategy for the robot answers was to always contradict the participant, except in the case where the items were completely unambiguous[5]. This required the operator to listen to the participant's answer and choose the appropriate answer each time. The answers were preprogrammed and available as a list on the GUI (Appendix 4). The order of the subtasks would randomly vary for each participant.

5 With a slight change with respect to the original paradigm [NAS 96], we decided to let the robot agree with the participant in the non-ambiguous questions, so that the participants could not be induced to think that the robot was always contradicting the human by default, or that the robot was faulty or lying in some way. The choice was made in the design phase of the experiment, after some tests with subjects in our laboratory, who reported to have the impression that the robot had a strategy in contradicting them in cases where the right answer was evident. With this little "trick", all our participants reported, after the experiments, that the robot had no strategy and was providing "real" answers.

#### 2.3.2.3.2. The social task

In the task designed to test the trust in the robot's social savvy, the participant had to answer three questions[6], by choosing which item between two is the most appropriate for a given context or situation (i.e. *at school, in a swimming pool, rainy day*). As for the functional task, two items were compared (Figures 2.6(a)–(c)), and the experimenter would ask each question ("*Which is the most important object at <context>: the <first item> or the <second item>?*") to the participant, then to the robot and then would ask the participant to confirm or not his/her choice (Figure 2.7). They (i.e. the participant and the robot) could disagree or agree, and there was no "right answer".

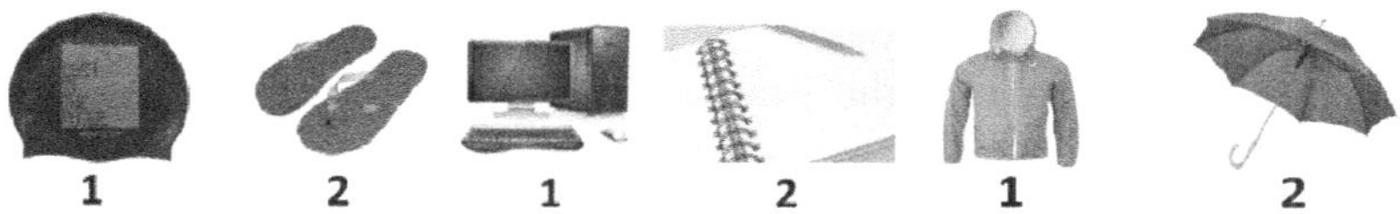


**Figure 2.6.** *(a–c) Social task: Q1: at school, which is the most important object: (1) the computer or (2) the notebook?; Q2: at the swimming pool, which is the most important object: (1) the bathing-cap or (2) the flip-flops?; Q3: in the rain, which is the most important object: (1) the K-way or (2) the umbrella?*

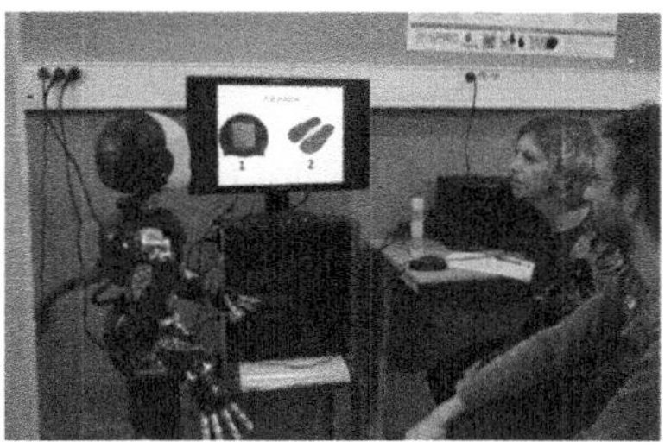

**Figure 2.7.** *Social task: the experimenter interrogates the participant and the robot*

---

6 We submitted to participants a limited number of questions (three) in the social tasks, for several reasons. One is to avoid the disengagement due to a long interaction with the robot, considering their frustration due to the fact that the robot is always contradicting them. Another reason was to avoid asking questions on matters where the participant could have had judgmental biases. Finally, as the functional task included three items (sound, color, weight), we chose to keep the same number of items for the social task in order to balance our experimental design.

While in the functional task the evaluations on the perceptual characteristics of the stimuli were based on objective measures (even if difficult to discern in the ambiguous or equal case), here, evaluation was essentially based on a subjective and personal judgment. Hence, the strategy chosen by the experimenter for the robot answers was to always contradict the participant as done by Nass *et al.* [NAS 96]. Again, this required the operator to listen to the participant's answer and choose the appropriate answer each time, formulated as "the first one" or "the second one".

#### 2.3.2.4. *Procedure and experimental design*

The present study was part of the Engagement During Human–Humanoid Interaction project[7]. The HRI protocol applied in this project was validated by the Ethical Valuation Council for Research on Health Issues[8].

Volunteers who took part in the experiment were required to fill in two questionnaires online at least 1 week before their visit to the Institute for Intelligent Systems and Robotics. These two questionnaires were (1) NARS [NOM 06] and (2) DFC scale [BUR 79].

On the day of the experiment, participants were welcomed by the researcher and informed about the overall procedure before signing an informed consent form granting us the use of all the recorded data for research purposes. Each participant was equipped with a Lavalier microphone to ensure a clear speech data collection.

Before the experiment, the participants had to watch a short video presenting the iCub, its body parts and some of its basic movements[9]. The video did not provide any information about the experiments. It was instrumental to make sure that the participants had a uniform

7 http://www.smart-labex.fr/index.php?perma=EDHHI.
8 IRB n.20135200001072.
9 This is a dissemination video from IIT showing the iCub, available on Youtube (http:// youtu.be/ZcTwO2dpX8A).

prior knowledge of the robot appearance (some participants may have seen the robot before on the media).

Fifty-six participants were all faced with the two tasks. One group (g1) was asked to imagine a collaborative scenario ($N$ = 21), while a second group (g2) was asked to imagine a competitive scenario ($N$ = 21). Finally, the control group (g3) was instructed to imagine a neutral scenario ($N$ = 14). The instructions related to the imagined scenarios varied according to the experimental conditions as follows. Instructions to elicit a collaborative (G1), competitive (G2) or neutral (G3) scenario are reported in Table 2.1. The five hypotheses were tested on the whole set of the participants.

| *Imagined scenario* | *English text* | *French text* |
|---|---|---|
| Competitive | Imagine that in 2 years you will be working with a robot to build some objects: you will be in competition. The one that will have built the best object will win a bonus.<br>Imagine this scenario in detail for a minute. | Imaginez que dans deux ans vous travaillerez avec un robot pour construire des objets : vous serez en compétition.<br>Celui qui aura construit le plus bel objet gagnera une prime.<br>Imaginez ce scénario de façon détaillé, pendant une minute. |
| Neutral | Imagine that in 2 years you will work with a robot. You will have to build some lovely objects.<br>Imagine this scenario in detail for a minute. | Imaginez que dans deux ans vous travaillerez avec un robot.<br>Vous devez construire de beaux objets.<br>Imaginez ce scénario de façon détaillé, pendant une minute. |
| Collaborative | Imagine that in 2 years you will work with a robot to build some objects: you will form a duo and will have to collaborate to build the object. As a duo you can win a bonus if the object is well built.<br>Imagine this scenario in detail for a minute. | Imaginez que dans deux ans vous travaillerez avec un robot pour construire des objets : vous formez un binôme et vous devez collaborer pour construire l'objet.<br>En tant que binôme vous pouvez gagner une prime si l'objet est bien construit.<br>Imaginez ce scénario de façon détaillé, pendant une minute. |

**Table 2.1.** *The instructions used for the imagined scenario of the human–robot collaboration*

The participant was then introduced to the robot. The experimenter did not present the experimental setup (e.g. show the location of the cameras) except showing the robot, and she/he did not provide any specific instruction to the participants about what to do or say and how to behave with the robot. Most importantly, she/he did not say anything about the way the robot was controlled: since the operator was hidden behind a wall, mixed with other students of the lab, the participant had no clue that someone else controlled the robot[10]. The robot was standing on its fixed pole, gently waving its hands and looking upright. It was not speaking.

Once the participant was standing and looking at the robot, s/he was free to do whatever s/he wanted: talk to the robot, touch it and so on. The experimenter took a seat on the right of the participant, in front of the robot and invited the participant to take the other seat.

The experimenter then provided verbal instructions for the experiment, consisting of two tasks: a functional evaluation task, aimed at assessing the trust in the robot's functional savvy, and a social evaluation task, aimed at assessing the trust in the robot's social savvy. The participants executed the two tasks in a random order.

Upon task completion, participants were asked to rate, on a 7-point scale (1 = *very competitive*; 7 = *very collaborative*) whether they recalled the imagined scenario as competitive or collaborative to ensure that they actually imagined the proposed scenario. The interaction task lasted on average 30 min per participant. The whole experiment took place in individual sessions in the experimental lab room of the iCub and it lasted on average 50 min for each participant. When the participant had finished, the experimenter thanked her/him and stored the collected data.

10 After the experiments, we asked the participants if they thought or had the impression that someone controlled the robot: all the participants thought that the robot was fully autonomous.

#### 2.3.2.5. *Data collection and analysis*

The participants and the robot responses were recorded on an individual sheet by the experimenter and could be additionally retrieved by the audio and video recordings.

Answers to the questions addressed during the functional and social tasks were used to create quantitative measures of participants' trust in the robot's functional and social savvy. The registered data consisted of the participants' conformation to, or disagreement with, the robot's answers.

A conformation score was calculated dividing the number of instances where the participant changed his/her answer to match the robot's answer (conformation) by the total number of instances where the robot's answer was in disagreement with the participant's first answer.

Responses to the two questionnaires were used to create quantitative measures of participants' DFC and attitudes toward social influence of the robot. Scores to the S2-NARS questionnaire were calculated in compliance with the method recommended by the authors [NOM 06]. The S2 score may range from 0 to 35. A high score indicates negative attitudes toward the social influence of robots. Scores to the DFC scale [BUR 79] were calculated according to the authors' method. The score may range from 0 to 140. A high score indicates a strong DFC. Over 56 participants, only 51 correctly filled in the DCF questionnaire. The results based on the DFC scale are obtained by retaining only these 51 participants. These three scores were used as dependent variables. As the functional and social conformation scores did not present normal score distributions, to carry statistical analysis we relied on non-parametric tests.

## 2.4. Results

### 2.4.1. *Do participants conform their answer more to iCub's answer in the functional task than in the social task? (H1)*

The conformation score obtained on the functional and social questions ranked from 0 (never conform) to 1 (always conform). In order to calculate to what extent participants trust robots during functional and social tasks, we have considered a score threshold of 0.5, that is the middle of the score scale, ranging from 0 to 1. We have thus esteemed that a conformation score higher than or equal to 0.5 in each of the two tasks reveals a participant's trust in the robot's savvy.

Descriptive analysis performed on the 840 (= 56 × (12 + 3)) participants' answers reveal that the average conformation score for functional task is 0.315 ($\sigma$ = 0.2), while the average conformation score for social task is 0.199 ($\sigma$ = 0.22). In both functional and social tasks, the mean scores are lower than 0.5 indicating as a whole that participants do not conform so easily to the robot answer. However, the Wilcoxon test for paired samples performed shows a significant difference ($V$ = 981; $P$ < 0.001) between these two scores. Participants tend to conform more to the robot's functional than social answers (Figure 2.8). Hence, our first hypothesis is confirmed: participants trust more in the iCub's functional savvy than in its social savvy.

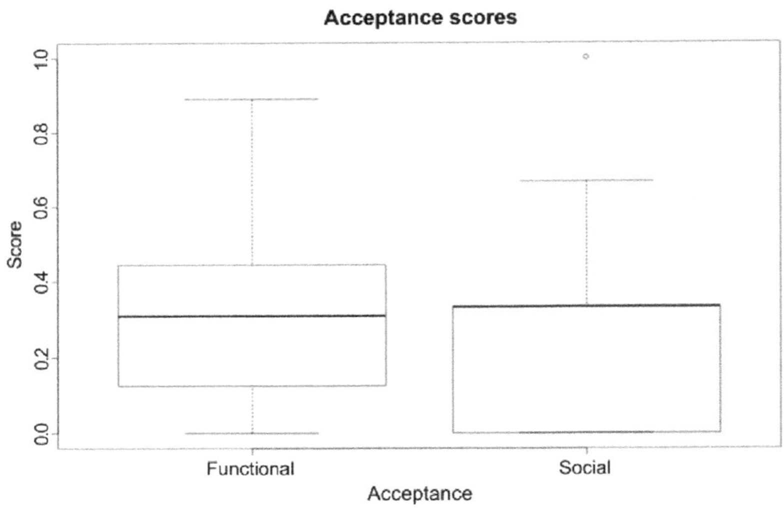


**Figure 2.8.** *Conformation score in the functional and social tasks for 56 participants*

### 2.4.2. *Do participants who conform to iCub in the social task also conform in the functional task? (H2)*

Among the 56 participants, 13 obtained a high functional conformation score (≥0.5) while only three participants showed a social conformation score higher than or equal to 0.5[11]. Table 2.2 outlines the conformation score of these three participants for both functional and social tasks.

| Participant | Conformation score | |
|---|---|---|
| | Functional task | Social task |
| #57 | 0.333 | 0.667 |
| #76 | 0.333 | 0.667 |
| #101 | 0.5 | 1.0 |

**Table 2.2.** *Conformation score in functional and social tasks of three participants who showed trust in social savvy*

Such a low number of participants showing a high conformation score in the social task does not allow us to carry out statistical tests. However, we may observe that two of these three participants also obtained a lower score for the functional task and one a score equal to 0.5, that is the threshold of low/high score. These results tend not to confirm our second hypothesis, i.e. participants do not trust social savvy uniquely, so that those who conform to the robot in the social task also would have conformed in the functional task.

### 2.4.3. *Does the imagined HRI scenario influence trust in iCub? (H3)*

To assess whether the two scenarios had an impact on participants' conformation, we performed a non-parametric ANOVA (Kruskal–

11 Our data show that the majority of our participants (92.8%) present a low conformation score in the social task revealing the following distribution: (always disagree) 0 (48.2%), 0.33 (44.6%), 0.5 (1.8%), 0.67 (3.6%) and 1 (1.8%) (always conform).

Wallis test). The main between-subject factor is the scenario condition (three levels: collaborative, competitive and neutral). The dependent variables are the functional and social conformation scores.

Results do not show any effect of the imagined HRI scenario on participants' conformation score in the functional task ($\chi^2$ = 1.69; ns, see Figure 2.9) or in the social task ($\chi^2$ = 1.63; ns, see Figure 2.10). Thus, our fourth hypothesis was not confirmed: imagining a collaborative interaction with the robot does not seem to increase trust in the robot.

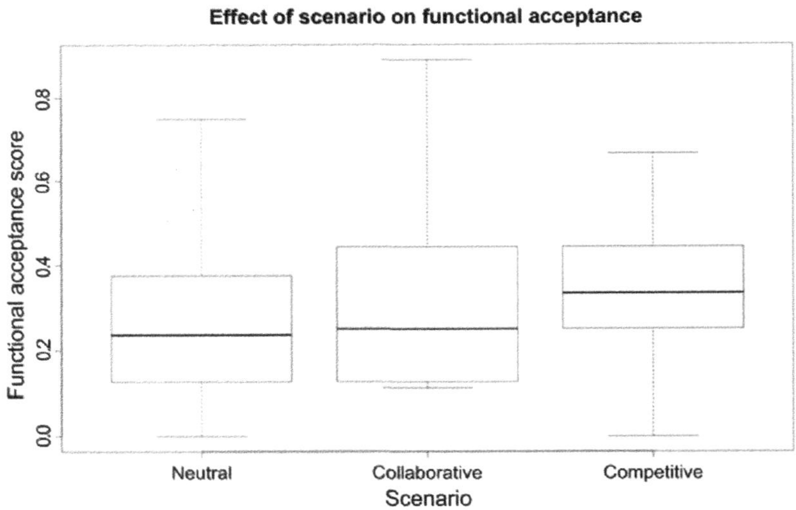


**Figure 2.9.** *Effect of the imagined HRI scenario on participants' conformation score in the functional task*

After the task, we checked whether participants remembered the type of scenario described by the experimenter (collaborative, competitive and neutral scenario). To do so, we asked the participants to rate, on a 7-point scale (1 = *very competitive*; 7 = *very collaborative*), whether they recalled the kind of scenario they had imagined. We carried out a non-parametric ANOVA with this score as a dependent variable and the imagined scenario as a factor. Results show a significant difference among the three imagined scenarios ($M_{competitive}$ = 3.57; $M_{neutral}$ = 4.76; $M_{collaborative}$ = 5.21; $\chi^2$ = 13.83; $P < 0.001$). Moreover, the Wilcoxon post hoc tests allow us to observe that there is a significant difference between competitive and neutral scenarios ($W$ = 325.5; $P < 0.001$) and between competitive and collaborative scenarios ($W$ = 245.5; $P < 0.001$), while no significant difference has been registered between collaborative and neutral scenarios ($W$ = 177; ns).

These results show that the participants in the competitive group tend to recall the scenario as less competitive than actually described in the instructions given by the experimenter. Therefore, our results showing no effect of the scenario on functional (Figure 2.9) and social (Figure 2.10) conformation should be taken carefully.

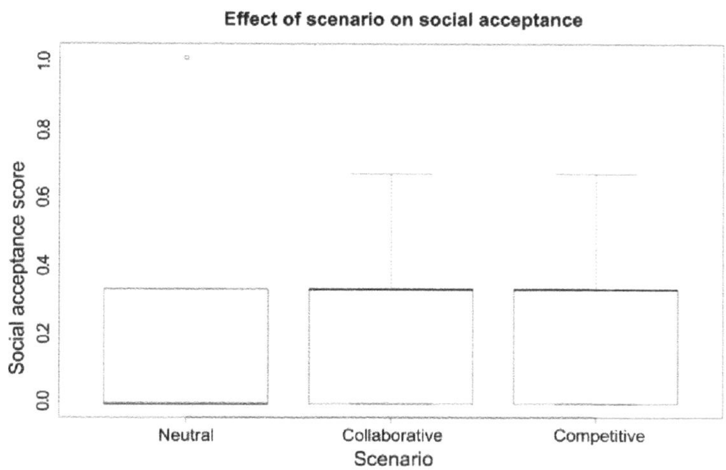


**Figure 2.10.** *Effect of the imagined HRI scenario on participants' conformation score in the social task*

### 2.4.4. *Is there a correlation between negative attitudes to the robot's social influence and the trust in the robot's social savvy? (H4)*

Descriptive analysis of the NARS-S2 (subscale 2 of NARS) scores shows that they follow a normal distribution ($W$ = 0.977; ns, Shapiro test): $M$ = 18.80; SD = 5.83. In order to identify a potential correlation between the conformation score and those of the NARS-S2, we performed a Spearman non-parametric test of correlation (Figure 2.11). Results do not show any significant correlation between the conformation score in the functional task and the score assessing the negative attitude toward the social influence of robots (NARS-S2) ($\rho$ = 0.127; ns) nor between the conformation score in the social task and the NARS-S2 attitude score ($\rho$ = 0.127; ns). No difference on this correlation was observed in each of the three imagined scenarios.

These results infirm the hypothesis that a negative attitude toward social influence of robots negatively correlates with the trust in the robot's social savvy expressed by the conformation score.

### 2.4.5. *Is there a correlation between the human DFC and the trust in the robot's functional and social savvy?*

Descriptive analysis of the DFC scores shows that they follow a normal distribution ($W$ = 0.964; ns; $M$ = 98.7; SD = 11.1). To identify a potential correlation between the conformation scores and the "desire for control" scores, we performed a non-parametric correlation Spearman test (Figure 2.12). Results do not show any correlation between the conformation scores in the functional task and the DFC score ($\rho$ = –0.086; ns), nor between the conformation scores in the social task and the DFC score ($\rho$ = 0.137; ns). No difference in this correlation was observed in each of the three imagined scenarios.

These results do not confirm the hypothesis that DFC negatively correlates with the trust in the robot's functional and social savvy expressed by the conformation score.

Finally, we observe a negative correlation between the DCF and the NARS S2 ($r^2 = -0.443$; $P < 0.005$), that is, the more participants have a high DFC, the less they show negative attitudes toward the social influence of the robot.

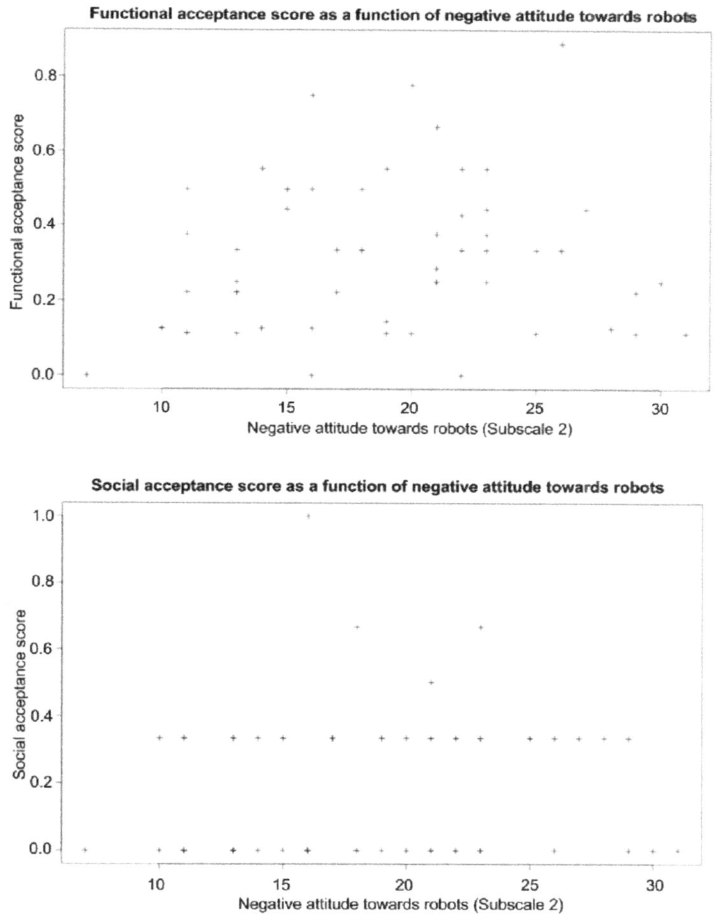


**Figure 2.11.** *Correlation matrix between the conformation score of the functional and social task and the scores of the S2-NARS*

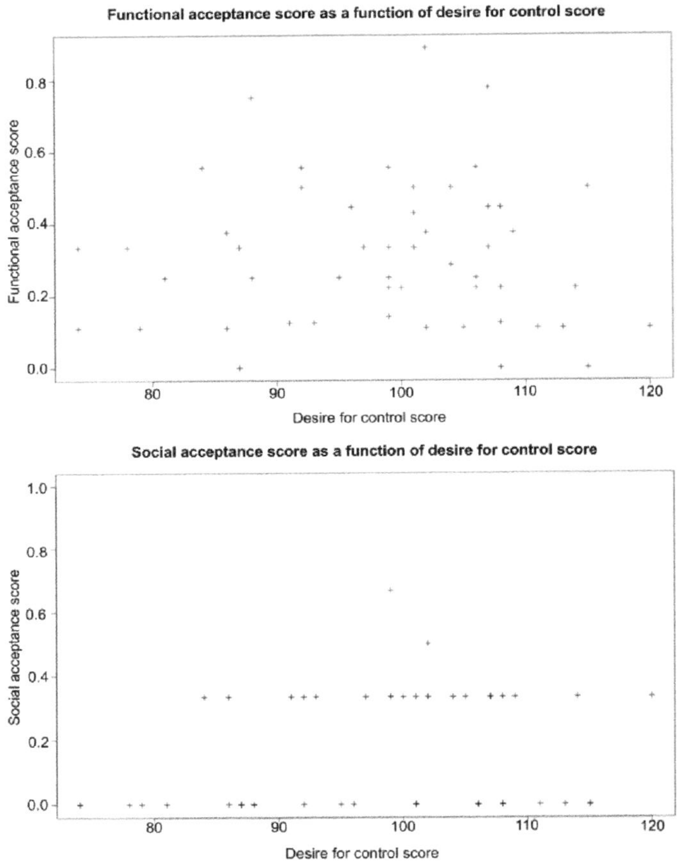


**Figure 2.12.** *Correlation matrix between the conformation score for functional and social task and those of the DFC*

## 2.5. Discussion

Based on the studies showing that the user's trust is a fundamental ingredient of HRI and can thus be employed as an indicator of robot acceptance [PAR 97, HAN 11, OSO 13, SCH 13], the present study aims at gaining insights into the human–robot trust dynamics that could affect the acceptance of robots in daily situations.

In particular, we were interested in investigating trust in functional savvy [SHA 09, HEE 10, FRI 14] as an indicator of users' acceptance of the robot as a functional agent, and trust in social savvy [WEI 09, YOU 09, DE 15] as an indicator of acceptance of the robot as a social agent.

While several empirical studies prove that people do trust computers social savvy during human–computer collaborative interaction [NAS 95, NAS 96], it is still not clear if this holds true for human–robot and particularly for human–humanoid interaction. Indeed, users consider functional savvy as indispensible for accepting the robot [KAP 05, BEE 11, HEE 009, WEI 09, FRI 14], social savvy is rather desirable [DAU 07, LOH 10] – but sometimes deceptive [DUF 93, COE 12].

Thus, the first purpose of our study was to find out whether users trust robots both when required to take decisions on functional issues and when required to do it on social issues. The media equation theory paradigm [NAS 00] provides a sound basis to investigate trust in technology by registering users' conformation or disagreement with computers.

Based on this experimental paradigm and on the research on robots' acceptance presented above, we hypothesized that trust in functional savvy should be more common than trust in social savvy and predicted that participants would have conformed to iCub answers more in functional than in social decision-making tasks.

Results show that in general participants do not easily conform to iCub answers. However, those who conformed mostly did it in the functional task. Hence, our first hypothesis was partially confirmed: participants tended to see robots as untrustworthy, or at least not trustworthy enough to take better decisions than humans, except when these decisions concern functional issues which require high-precision technical skills to discriminate a specific perceptual characteristic of the stimuli. In this case, participants relied on the robot's functional savvy and accepted more often that iCub answer determines their final decision. The exam of verbal registrations of participants' answers confirms this reliance: several participants reported that they preferred to rely on iCub's perception than on their own perception. Furthermore, these registrations let us to remark that participants who did not conform in the functional task motivated their lack of conformation referring to their expertise in a specific domain. For example, one participant said that she was a musician, and so he relied more on her own ability to detect sounds than the robot's ability to detect sounds.

The differences observed among the participants regarding their trust in the robot's functional and social savvy could then be explained by the fact that, since functional savvy is based on an objective knowledge and on highly technical skills, which participants tacitly ascribe to robots as a core knowledge [KAP 05, BEE 09, HEE 09], the robot is more easily accepted as a reliable tool to help with functional duties. Although social savvy is based on a common-sense knowledge that is usually acquired through repeated experience in social contexts [GUI 99], it can be implausible for users to assume that such savvy is instilled in the robot like a built-in setting, or to believe that the robot has a life history behind it, made up of social interactions and situations. Consequently, the robot is hardly accepted as a reliable partner in social tasks, at least at the current stage of the introduction of robots in social contexts. To this concern, commentaries of participants retrieved in the verbal registration witness that some participants raised his/her concern about the credibility of the robot when answering social questions. For example, one participant remarked that the robot had never gone to the swimming pool, so it could not know which was the most appropriate item in that context. Similarly, several participants remarked that the robot had never been out in the rain, so it could not know which was the most appropriate item in that context either.

Another possible explanation is that while the validity of decisions taken to achieve functional tasks can be corroborated by the physical reality (i.e. the impression of "good or wrong estimation" can be confirmed or not by scientific evidence, see [KAH 07] for a review), the validity of decisions taken to achieve social tasks (i.e. the judgment of "good or wrong doing") can be validated only intersubjectively [NIC 04]. Therefore, while in functional tasks participants' decision-making is achieved with the awareness that they are not the "final judge" of the perceived reality, so that they can leave the decision to the robot, social tasks are more likely to arouse a persuasive behavior, which leads participant to give more importance to their personal judgments and to defend them intersubjectively (i.e. in front of iCub) as the best decision in a given context. In fact, while performing the social task several participants commented that they wished to express their own opinion even when knowing that robot's answer was the correct one.

Moreover, looking at this result in the light of those obtained by Nass *et al.* [NAS 00], we can point out that, despite human-like robots are increasingly conceived and perceived as social technologies [BAR 07b], we did not registered a comparable or more significant level of social trust with relation to robots than the one registered by media equation theory studies with relation to computers. These results suggest that, currently, users trust more the computer social savvy than the robot's social savvy. If at a first glance this may seem counterintuitive, we should consider that while we today have little familiarity with robots, our usage of computers is frequent and fluent. It is then possible that we trust computers more because we are already acquainted with them, and thus we tend to believe that no additional effort is needed to use them for social tasks. Furthermore, while a computer is characterized by powerful computation functionalities that can support decisions, the embodied and behavioral nature of robots potentially enable them to transform computations and decisions into actions. It is thus reasonable to think that a robots' social savvy may be seen as having more straightforward and intrusive consequences than a computers' social savvy, and this can cause a distrust bias as a manifestation of the anxiety toward the robots during social tasks. However, contrary to other experimental paradigms where participants are asked to take decisions that imply concrete consequences (see [KHA 07]) or that confront them with a "moral dilemma" (e.g. [MAL 15]), in our study the decisions taken by participants do not have any consequential effect on real actions or on morality. Therefore to validate this interpretation of the results, in future research we shall introduce a decision-making task whose objective is not only to provide a good answer but also to care about the results of this answer.

Interestingly, not only is trust in robots different from trust in computers, but also from trust in humans [ROT 71, MAY 95, BIL 12]. It has been proven that although users are rather inclined to project a social life on robots, i.e. they tend to interpret robot behavior as they do with humans or animals by attributing animacy and social skills to them [SEV 10], they do not attribute free will and intentionality to robots, i.e. they do not believe that robots are capable of voluntary choice and social judgment [MON 14]. It is thus possible that in our study participants did not conform to the iCub because they did not attribute it either the capability of voluntary choice nor the common-sense knowledge or social judgment, which is underpinned by this capability. The resulting portrait of the robot is thus one of a social but not intentional agent and maybe this incongruous status is what makes the robot unique.

In this sense, trust in robots may present specificities with relation to trust in computers and humans, so that these specificities can be further investigated in future research by comparative experimental conditions involving human–computer interaction, HRI and human–human interaction. In particular, given the child-like aspect of iCub, it would be interesting to compare interactions between two children versus interaction between a child and iCub.

To continue on our findings, following the proposal of Young *et al.* [YOU 09] that trust in functional savvy is a prerequisite for trust in social savvy, our second hypothesis was that users would not trust the social savvy uniquely. Consequently, we predicted that participants conforming to the iCub's answers in the social task would also conform in the functional task. Contrary to our hypothesis and to the current literature [YOU 09], results show that the very limited number of participants conforming in the social task conformed less in functional task.

This result indicates that the few participants who believed that social savvy is susceptible to be an intrinsically reliable ability of the robot do not believe the same for functional savvy. This minority does not base trust in social savvy on trust in functional savvy, and on the contrary seems to assume that a social robot that has poor technical skills and little objective knowledge would not necessarily be an untrustworthy social robot.

Globally, this first set of results leads us to conclude that, over the limited portion of participants who trusted in the robot, a restrained number considered that trust in functional savvy excludes trust in social savvy, and an even more restrained number of users who considered that trust in social savvy excludes trust in functional savvy. This mutual incompatibility between trust in functional and social savvy may contribute to a deeper understanding of users' trust and distrust behaviors: if these behaviors have indeed been observed in previous studies with relation to a robot's level of autonomy [HEE 10, SCH 13], adaptability [KAM 13] and quality of movements [BRU 13], our study put the accent on the nature of the task itself, since this latter concretely instantiates the double nature of the humanoid robots, which are conceived to be both functional and social devices.

With regard to the DFC, the negative attitude toward robots' social influence, and the simulated interaction scenario with a robot, the results only partially confirmed our hypothesis as well as the discussed literature.

More in detail, results concerning the influence of the DFC on participants' conformation indicate that, whether participants showed a high or low DFC, they mainly tended not to conform to iCub. This confirms previous studies where users generally preferred to be in control of the robot [GIL 04, MAR 04, SYR 07, OKI 12, KAM 13,

KOA 14]. With regard to the 13 participants conforming to iCub in the functional task, and to the three participants conforming to iCub in the social task, we might ascribe this behavior to the fact that the young age of iCub, whose appearance was that one of a 4 year old robot, aroused a tolerant behavior, so that these participants did not wish to contradict iCub just like sometimes parents or teachers find it difficult to contradict a child.

Concerning results on the influence of the attitude toward robots, results show that the participants' trusting behaviors were independent of the negative attitude toward social influence of the robot: whether they feared being influenced by the robot or not, participants did not conform their decisions to those of the robot, except for the minority who conformed during functional tasks and for the even more restrained minority who conformed during social tasks. This could mean that distrust is itself a deeply rooted and widely diffused attitude toward robot, more than specific fears about robots, like the fear of being influenced by them. Hence, even when these fears are overcome, if they are, still psychologists and engineers will have to deal with a generalized and substantial level of distrust amongst users. Further analysis have also shown that the more participants have an elevated DFC (score at the DFC), the less they fear to be influenced by robots (score at the NARS-S2). These participants seem thus not to be concerned by the possibility that a robot could determine their decisions as they probably assume that they will still be in control of their functional and social tasks and they would not need to rely exclusively on robots. This result seems to indicate that participants with strong DFC interpret trust mainly as reliance – with reliance being one of the components of trust [BIL 12] – and that idea of being in the condition to rely on a robot makes them unwilling to use the robot. Correlation between desire to control and negative attitudes toward a robot could thus turn out to be revealing of intention to use the robot. Intention to use the robot predates any eventual trust behavior, and thus supplies an early evidence of a tendency to resistance, rather than to acceptance. Prospective developments of the present study should thus identify the participants' levels of intention of use as a preliminary predictor of trust behavior. Moreover, since this study mainly focuses on robots' functional and social acceptance, it shall be useful to additionally assess participants' regular use of technology, by asking them if they normally use social technologies (e.g. social networks, videogames, etc.), and whether they use functional technologies for social purposes (e.g. using an office computer for chatting, watching a movie, etc.) as well as social technologies for functional purposes (e.g. using social networks to retrieve useful information, selling objects and looking for a job). These pieces of information may help to predict trust behaviors in future research.

Finally, results of the influence of the imagined scenario on trust behavior seem to infirm our hypothesis that imagining a collaborative interaction scenario would have determined a higher level of trust, and while this is in contrast with the findings in the study of Kuchenbrandt *et al.* [KUC 12] witnessing that robots acceptance increases when users imagine a collaborative scenario, this is in line with the more recent study by Wullenkord *et al.* [WUL 14] where participants who had imagined contact with a robot did not report more positive attitudes toward robots. However, in the specific case of our study, the weak influence of the imagined scenario on participants' level of trust can be explained by the fact that participants had difficulties in imagining the proposed interaction scenario: apparently, participants who had been asked to imagine a competitive scenario could hardly recall that they imagined it. This can be due to the fact that while in the imagined scenario participants thought about an abstract non-specified robot, in the real interaction they were confronted with a specific robot that has a child-like appearance. Again, the appearance of iCub could have diminished the strength of the competitive scenario and enhanced the strength of collaborative scenario in their memory. Therefore, in future research we shall vary the physical appearance of the robot to validate our hypothesis on the influence of imagined scenario on users' trust.

## 2.6. Conclusions

This study shows that robot acceptance is a complex dynamic characterized by a prevailing distrust in robots, and where the limited trust behaviors that can be observed among users are significantly correlated to the nature of the task at hand. In particular, robots seem to be more easily accepted in functional than in social tasks. This is witnessed by the evidence that when confronted with tasks requiring decisions about functional issues, users trust robots more than they do when confronted with tasks requiring decision on social issues. Moreover, the minority of users who trust robots on social issues show a significant distrust in robots on functional issues.

These results do not allow us to understand whether the observed distrust and trust behaviors depend on the fact that, despite today the robot's vocation being at the same time a functional and social technology, users still consider robots as "socially ignorant" [YOU 09] or rather on the fact that trust in functional savvy and trust in social savvy are mutually exclusive because they are based upon different kinds of knowledge and skills (scientific objective knowledge and technical skills vs. subjective common-sense knowledge and adaptive skills based on social norms).

However, the general distrust and the different behavior of participants in the two tasks suggest that trust in robots cannot be assimilated to mechanisms that are typical of trust in computers neither to those that are typical of trust in humans.

Furthermore, the three observed HRI factors (DFC, negative attitudes toward social influence of robots and imagined interaction scenario) seem not to have an influence on trust behaviors in terms of conformation of users' decisions to robot decisions.

Nonetheless, we registered a significant correlation between DFC and negative attitudes toward robot. This correlation indicates that the more participants wish to have control on their life situations, the less they fear being influenced by a robot, and this suggests that users with control-seeker profiles do not wish to use robots as far as they would have to rely on them. Consequently, we argued that correlation between DFC and negative attitudes toward robots could help in quantifying individuals' intention of use. Furthermore, we proposed that the intention of use can be an interesting predictor of trust for participants whose strong DFC implies that they reduce the complexity of trust to one of its components, which is reliance (e.g. dependence from robots), and thus tend to resists robots rather than accept them. Additionally, we have proposed that users' habits with concerns to functional and social technologies can also be investigated as predictors of trust behavior.

Finally, from a methodological point of view, our study is an attempt to establish a conceptual bond among pivotal issues that are being recurrently used in the HRI communities such as "social and functional acceptance", "robots' savvy" and "trust". These issues, though being recurrently used in the HRI communities, still suffer from some fuzziness since they have been inherited from other domains, namely ergonomics and social psychology, and need to be adapted to social robotics in order to obtain a more profound understanding of users' acceptance of robots.

# 3

# Learning by Robotics: The Impact of Educational Robots on Learning

## 3.1. Combining RBI and inquiry-based science

Formerly limited to academia [SKL 03, PET 04, BAR 07, WIL 07, NUG 10], robotic-based instruction (RBI) today crosses the doorstep of kindergarten [TAN 05, TEJ 06, SUL 15], primary school [BER 02, KAN 03, MIT 08, RIB 08, CHA 10, MCD 12, KIM 14] and secondary school [ROB 05, HUS 06, LIN 07, WHI 07, BAR 09, CHO 15].

The reasons of such a pervasive diffusion of RBI in educational contexts, under the form of modular activities, short-term and long-term projects based on constructible and programmable robotic kits (Lego Mindstorms®, Lego WeDo®, etc.), can be found in the educators expectations toward a technology that has been defined, from its birth, as an *object to think with* [RES 96] or a *mindtool* [MIK 06] as well as a *chameleonic tool*, which can adapt to different pedagogical objectives [STR 92]. In this sense, the attractiveness of robotic kits resides in their "low floor", i.e. robotic kits offer a facilitated approach to robotics, "high ceiling", i.e. they allow a diversity of projects, and "wide walls", i.e. they encourages different learning styles [PAP 81].

Robotic kits are normally composed of mechanic and electronic elements that are controlled by dedicated interfaces where students program robots' behavior through high-level language (i.e. a programming language that is closer to the human natural language than to machine language), that can be verbal (Logo®, Scratch®, etc.) or graphic (LEGO NXT®, Labview). Since robotic kits have succeeded kits containing exclusively bricks (e.g. traditional LEGO® kits) and kits containing exclusively mechanisms (e.g. Fischertechnik®), they are also called "third generation technology".

Concrete applications of RBI have been distinguished into three types [ALI 13]: (1) theme-based curriculum approach, where different curriculum areas are integrated around a specific topic and studied mostly through inquiry; (2) project-based approach, where students work on a project to explore real-world problems; (3) goal-oriented approach, where students compete in a robotics challenge (e.g. FIRST LEGO® League[1], RoboCupJunior[2], Trophée de robotique[3] in France, etc. [LUN 00]).

These applications of RBI affect both domain-specific knowledge and transversal competencies. That is, their usage has been suggested to improve students' results in science, technology, engineering and math (STEM) areas [HUS 06, WIL 07, MIT 08, BAR 09, NUG 10], life sciences [WHI 07, BRE 10], music and art [RES 91, RIB 08, RUS 08], but also to foster transversal competencies such as problem solving, scientific process skills, team working, planning, self-correction, self-confidence, creativity, etc., that can be transferred to other domains, beyond robotics [DEN 01, SKL 03, CHA 07, LIN 07, SUL 08].

However, as a recent survey points out, the empirical evidence to support the effectiveness of educational robotics (ER) is rather limited [BEN 12]. As this researcher remarks, most of the literature on the use

1 http://www.firstlegoleague.org.
2 http://www.robocupjunior.org.
3 http://www.planete-sciences.org/robot.

of robotics in education is descriptive or anecdotical, based on the reports of teachers [LAU 99, CAC 03, PET 04, FOR 06], while studies based on rigorous experimental design, and in particular studies with control groups, are quite rare. Consequently, it is difficult to isolate the factors that contribute to the success of robotics projects. Benitti [BEN 12] thus concludes that, even if robotics has much potential to assist in teaching, the gain in learning by students is not guaranteed by the simple use of robotic kits, as there are several factors that may favor or prevent this gain, such as guidance on how robotics activities can relate to lesson content [NUG 08], the fact that the proposed task should be realistic and relevant for students [LIN 07] and, more importantly, the combination of the robotics environment with specific pedagogical approaches [SUL 08].

In this chapter, we would like to focus precisely on this: the pedagogical approach as a key factor of RBI success. As several studies have pointed out, technology alone cannot automatically enhance the quality of teaching-learning dynamics, unless the use of technology is "orchestrated" [TRO 05] or "tamed" [CHA 02] through a suitable pedagogical approach that allows, on the one hand, to transmit the lesson content through new technologies and, on the other hand, to appropriately integrate these new technologies as learning tools.

To this end, the literature shows that studies using robotic kits straightforwardly (i.e. without a specific pedagogical approach) have registered improvements that are limited to specific facets of learning and populations of students who already had a good level of results before using robotic kits; while the few studies where robotics kits were coupled with a pedagogical approach have registered more significant and inclusive results. For example, Lindh *et al.* [LIN 07] found that after having used robotic kits, students showed limited improvements in problem-solving skills and these improvements occured only for students without particular difficulties. Hussain *et al.* [HUS 06] remarked that interest in robotic kits is typical of students who were already high performing in mathematics. Nugent

[NUG 08] reported students' progress specifically on STEM competencies [BAR 07a]. Mitnick *et al.* [MIT 08] registered improvements related to collaborative attitudes. Finally, in a previous study, we found that using robotic kits helps educators to identify the learning strategies employed by children when facing unfamiliar educational technologies [GAU 12b].

By contrast, Sullivan [SUL 08] pointed out that when RBI is coupled with a pedagogical approach – that includes both direct instructions (short lectures, software, demonstrations and lab assignments) and open-ended, extended inquiry (i.e. students working in pairs to solve problems posed as programming and design challenges) – pupils show learning improvements on a wider range of knowledge and skills: their utilization of science literacy-based thinking as well as science processing skills is enhanced and their understanding of systems increases.

A pedagogical approach for robotics is a recent field of investigation that calls for a "shift from technology to pedagogy" [ALI 13]. RBI is inherently concerned by the wide-range evolution of education from instructionist to constructivist approach [PAP 80], i.e. from teaching-learning dynamics mainly directed by the teacher who transmits knowledge and propose exercises, to an approach that is rather discovery oriented, experiential, self-directive, and which alternates phases of concrete activities and abstract thinking [PIA 72]. With relation to RBI, this new trend has taken the form of "costructionism" and "guided constructivism" [ACK 01, ALI 09], where students' projects are carried out under the guidance of the teacher, who scaffold class learning by recalling objectives, prompting collaboration and generalizing good practices. Inheriting its theoretical fundaments from constructivism [KEL 55, PIA 72, GAT 03], constructionism and guided constructivism highlight that learning is not a simple accumulation of knowledge, but rather the result of the active engagement of the learner who constructs and co-constructs [VYG 62] knowledge from the concrete, situated and social experience [SAL 96].

Through time, RBI has also been claimed to fit a number of pedagogical approaches such as *discovery learning* [BRU 61, MAY 02b], *serious game* [WOU 09], *project-based learning* [THO 00, TRE 97], *problem-based learning* [GAL 92], *learning by doing* or *hands-on learning* [SCH 00] and *teaching and learning by design* [BER 02].

Each of these approaches emphasizes a specific aspect of learning. *Serious games* put the accent on playful and immersive atmosphere, as a mean to reduce students' apprehension for negative results, give them the feeling to be in a "flow", and thus to learn effortlessness and progress faster. *Project-based learning* implies lessons that are coherently articulated around a theme of authentic interest for children, rather than traditional sequential lessons. *Problem-based learning* replaces school-like exercises by realistic cases of investigation. *Learning by doing* encourages practical activity to experience knowledge in action rather than passively receive knowledge from manuals and educators. Finally, *teaching and learning by design* proposes that taking in charge the customization of teaching and learning material rather than using already made material improves the quality of teaching and learning.

But learning is a composite process of knowledge, competencies and skills acquisition that holistically involves all the above-mentioned aspects. From a psychological point of view, these aspects can be ascribed to four main dimensions: cognitive, affective, social and meta-cognitive [SHO 89].

Therefore, in order for RBI to have an impact on learning, it should be combined with a pedagogical approach that possibly covers the whole four dimensions, is compatible with a robotic-based environment and ties to official school programs. As Alimisis [ALI 13] recalls, *the fundamental issue is not the robot, but the alignment of technology with a sound theory of learning and with curriculum*.

The approach on which we focus in this book is IBSE (inquiry-based science education). IBSE is a pedagogical frame where pupils are confronted with open interrogatives or challenges, whose answers and solutions imply the acquisition of empirical, collaborative and transferable learning [BEL 10]. There are at least three reasons for focusing on this approach among the approaches that have been proposed to scaffold robotic activities.

First, IBSE is a comprehensive approach that results from gathering the principal issues raised within each of the above-mentioned constructivist approaches [KOL 03, HAE 04] and takes care of learning holistically by targeting the development of its cognitive, affective, social and meta-cognitive aspects [RIE 16].

Second, because this approach has already proved its efficacy at primary and secondary levels in increasing students' and teacher motivation [ROC 07], stimulating students' curiosity about the world [ENG 08, BAR 09], fostering their understanding of how scientists build scientific knowledge [ESH 06] and enhancing their awareness of the uncertainty that exists within the scientific enterprise [OSB 08].

Third, because IBSE is the quintessence of the renewal of the National Primary Curriculum in France, which will definitively be applied in 2016[4]: such renewal implies a passage from a science curriculum that was mainly focused on acquisition of basilar knowledge and competencies (*compute, solve mathematical and geometrical problem, organize data through tables and graphs, using the computer to search for information and elaborate documents, etc.*) to a curriculum that is strongly inquiry oriented and includes methods, processes and tools of sciences, digital know-how, engineering, history of scientific and technology ideas and enterprises and student-driven projects. This renewal is in line with a world-wide change in educational trends, "a reversal of school science-teaching pedagogy

4 For a prospective synthesis of the 2016 Primary Curriculum, see http://www.education.gouv.fr/pid25535/bulletin_officiel.html?cid_bo=87834#socle_commun.

from mainly deductive to inquiry based methods that provide the means to increase interest in science" [ROC 07].

## 3.2. IBSE and the four dimensions of learning

At the heart of the IBSE approach, there is the idea that learners are a single thing with the process of learning: they are committed in their education, whose achievement is not simply a duty toward teachers or parents, but toward themselves and toward the other learners of the class. This commitment is undertaken by carrying an inquiry [WHE 00, CUE 05, ZER 14].

Generally defined as "the dynamic process of being open to wonder and puzzlements and coming to know and understand the world" [GAL 04], inquiry is articulated by projects or activities that are negotiated with the class, developed by the students and facilitated by the teacher.

When applied to sciences, inquiry is defined as "the process of posing questions and investigating them with empirical data, either through direct manipulation of variables via experiments or by constructing comparisons using existing data sets" [QUI 04, p. 341].

More precisely, Quintana *et al.* [QUI 04] divide the processes of inquiry into three broad categories: *sense making* that involves fundamental operation like hypothesis making or data analysis, *process management* that consists of the strategies applied to control the inquiry process and *articulation and reflection* that includes constructive, evaluative and articulating processes. These processes are further detailed by Bell [BEL 10] who identifies nine activities of inquiry: *orienting and asking questions*, *hypothesis generation*, *planning*, *investigation*, *analysis and interpretation*, *model*, *conclusion and evaluation*, *communication* and *prediction*. The last two activities of inquiry, as Bell highlights, often lead to new questions, so that inquiry can be seen as a cycle [WIN 04, SCH 05] or an iterated sequence: "Question–Predict–Experiment–Model–Apply" [WHI 98]. Importantly, the inquiry process is modeled by graphic schemas that are circulated among students in order to increase their awareness of the steps of the inquiry and allow them to more easily identify, which knowledge and competencies are required for a given step [WHI 98].

The mentioned works provide a general definition of IBSE and of the way an IBSE lesson should be structured. Moving from these definitions, the EU project Pri-Sci-Net[5], whose main objective is to bring inquiry at a primary level by developing IBSE activities and networking those schools that adopt this approach [GAT 16], has formulated a set of principles that operationalize theoretical research on IBSE pedagogy by describing the role of teacher and students in IBSE, along with the core learning objectives of a IBSE lesson:

– Principle I: Active engagement of children in the learning process with emphasis on observations as a point of departure.

– Principle II: Authentic and problem-based learning activities where the correctness of answers is evaluated only with respect to the available evidence and getting to a correct answer may not be the point.

– Principle III: Self-regulated learning, social interaction and collaboration are emphasized.

– Principle IV: Emergent student autonomy, discursive argumentation and communication with peers (talking science) are encouraged.

– Principle V: The teacher adopts the role of a facilitator and represents an example of an inquiring person. The teacher does not function, in the eyes of the children, as the sole bearer of expert knowledge.

5 European Project Pri-Sci-Net: networking Primary Science Educators as a means to provide training and professional development in inquiry-based teaching (FP7-SIS-CAPACITIES), see: http://www.prisci.net/.

– Principle VI: Assessment has a formative role in providing feedback during the teaching and learning process.

Such principles have been co-built by the Pri-Sci-Net consortium to foster authentic, engaging, collaborative and self-regulated learning across Europe. They are meant to reflect the composite nature of learning, in which the cognitive, affective, social, and meta-cognitive dimensions are equally targeted.

### 3.2.1. *The cognitive dimension*

Broadly speaking, cognitive processing consists in the thinking activities that learners use to process learning content. They lead to learning results in terms of knowledge, understanding and skills. Examples are: looking for relations among parts of the subject matter (relating), distinguishing main and minor points (selecting), thinking of examples (concretizing) and looking for applications (applying) [VER 96].

In IBSE, the cognitive activities at play are mainly related to the problem-solving process [HME 04], so "applying" covers an important part of an IBSE lesson, while "relating" and "selecting" are important to understand the problem as a whole and compare differences and similarities in the tests' results.

In more detail, according to Anderson and colleagues [AND 01], children make use of particular types of cognitive processes when approaching a problem-solving task in an educational environment: *remembering*, which means retrieving knowledge from long-term memory; *understanding*, which is constructing meaning from instructional messages; *applying*, which is carrying out a procedure in a given situation; *analyzing*, which is breaking a whole into parts and apprehend their relation; *evaluating*, making judgments based on criteria and standards; *creating*, by putting elements together to form a coherent or functional whole [AND 01, p. 68]. These processes result in four categories of knowledge: factual knowledge, relating to the basic elements needed to know about a discipline or to solve problems in it (i.e. knowledge of terminology and specific facts); declarative knowledge, concerning the interrelations among the basic elements within a larger structure that enables them to function together (i.e. concepts, categories, principles and models for this discipline); procedural knowledge, constituted of the procedures, techniques and methods as well as the criteria for using them; and meta-cognitive knowledge, namely the knowledge of a task's objective, and of one's own knowledge and actions to be carried to achieve the objectives [FLA 79], which will be further detailed in the following chapters. Together, declarative and conceptual knowledge represent knowledge about "what", while procedural and meta-cognitive knowledge represent knowledge about "how".

IBSE is considered to foster the cognitive processes that support the acquisition of these four types of knowledge. Procedural understanding includes particular process skills such as questioning, observation, prediction and relates to complex thinking skills required to collect reliable data and draw and evaluate conclusions from evidence obtained [HAR 00, SAD 09]. Factual and declarative understanding involves the shift from spontaneous concepts or naïve knowledge to scientific concepts [VYG 62]: spontaneous concepts are progressively restructured as students are exposed to new experiences, facts and ideas, which challenge their preexisting beliefs and let them acquire new terminology [MIN 10]. Meta-cognitive understanding is mainly related to awareness of the process of inquiry and its main steps, as well as to transfer of the acquired inquiry skills to carry a new inquiry [WHI 98].

Accordingly, the pedagogical activities of the Pri-Sci-Net project have been conceived including the following phases: *engage*, *inquiry* and *evaluate*. In the engage phase, the teacher invites pupils to recall their pre-existing knowledge on the subject of the inquiry and to pose questions. The result of this phase is the definition of the main objective and hypothesis of their investigation. In the inquiry phase, pupils are called to observe, collect data and relate evidence and explanation. In the evaluation phase, pupils analyze the relation found between evidence and explanation, make inferences, establish first conclusions and eventually propose new interrogatives and predictions.

In our third experimental study (section 3.3), the cognitive dimension was investigated by questioning pupils about their ability to: (1) retrieve their preexisting knowledge, (2) correct their preexisting knowledge and (3) acquire new knowledge and skills.

### 3.2.2. *The affective dimension*

"Affective activities are directed at coping with the feelings that arise during learning and lead to an emotional state that may positively, neutrally, or negatively, affect the progression of a learning process" [VER 96]. These activities have been identified by Vermunt as motivating, concentrating, judging oneself, appraising, exerting effort and generating emotions.

The inquiry process brings with it various feelings, including enthusiasm, apprehension, frustration and excitement [ALB 04]. In a previous work [RIE 16], we defined the affective dimension focusing on two aspects: enjoyment and engagement. Enjoyment refers to sense of curiosity and excitement that often characterizes the beginning of the inquiry. Engagement refers to the sense of ownership of learning, which is experienced when being immersed in the challenge of the project and overcoming the encountered obstacles, conflicts and frustrations. Enjoyment and engagement are fundamental for self-confidence [BLA 07], the desire to participate as well as for sustained motivation [FRI 13]. As in IBSE, the learning outcome is not the low or high score attributed by the teachers, but the achievement of the project in which students are involved as individuals and as a group: students indirectly take in charge the responsibility of their learning by taking in charge the responsibility of the project [RIE 16]. Indeed, the most successful curriculum inquiry projects emerge from topics that are of personal interest to the students [WIG 98].

In our third study (section 3.3), the affective dimension is thus investigated by evaluating pupils': (1) levels of appreciation of the project (enjoyment) and (2) desire to continue or carry future projects (engagement).

### 3.2.3. *The social dimension*

Learning is intrinsically social [VYG 78]. Salomon and Perkins [SAL 98] distinguish six meanings of social learning:

1) Active social mediation of individual learning, where a facilitating agent (teacher, peer or group) and the primary learner form a joint system, the former helping the latter to achieve critical conditions of learning. As the authors remark, traditional instruction inevitably involves a certain amount of social mediated individual learning.

2) Social mediation as participatory knowledge construction, where individuals learn in a situated system, in which the interaction serves as the socially shared vehicle of thought. Here, the learning products of the system, jointly constructed as they are, are distributed over the entire social system rather than kept by the individual learner.

3) Social mediation by cultural scaffolding, where the learners enter into an intellectual partnership helped by artifacts and tools to share information, symbols, conjectures, etc.

4) Social entity as a learning system, where the focus falls on a collective agency that acquires knowledge, understanding, or skills, and where it is pointless for an individual to perform alone (e.g. a sports team).

5) Learning to be a social learner. This concerns learning to learn, the ultimate goal of learning. Declined with relation to social learning, learning to learn consists of learning when to ask questions and help and how to enter into reciprocal learning relationships. In this sense, learning to be a social learner means to acquire new ways to capitalize on the social surroundings.

6) Learning social content. Social content includes issues such as how to collaborate with others by expressing one's own point of view, respecting others' points of view, and taking decisions together.

However, as Salomon and Perkins [SAL 98] remark, for a long time there have been two separated conceptions of learning. On the one hand, there was a conception of learning as an individual process, emphasizing the individual acquisition of knowledge and skills as "transferable commodities". On the other hand, there was a sociocultural conception of learning as a participatory process of active knowledge construction emphasizing context, interaction and situatedness. This second conception comes from the conviction that social learning lets those cognitive conflicts emerge whose process of solutions is pivotal to shape students' sociocognitive development [PIA 52].

Within IBSE, these two conceptions of learning are no more separated. Inquiry often incorporates collaboration and engagement of participants in a common endeavor [DIL 99]. This implies that students have to communicate (to share information, ideas, argue their point of view and present their project) and coordinate their work in order to achieve their common project [DRI 94].

In our third study (section 3.3), the social dimension is investigated by assessing pupils' levels of communication during the main phases of the project: (1) within their group, (2) among the class groups as well as by asking them to (3) evaluate the quality of the group work.

### 3.2.4. *The meta-cognitive dimension*

Meta-cognition refers to awareness of one's capacity to learn, the nature of what is to be learned or task demands and actions needed to accomplish tasks [FLA 79, AND 01]. Meta-cognition thus implies the use of process management strategies [QUI 04] or regulative processes [DE 05] that have been identified by Vermunt [VER 96] as orienting, planning, monitoring whether the learning process proceeds as planned, testing, diagnosing the cause of difficulties, adjusting, evaluating and reflecting. These strategies or processes help to organize and internalize the learning process, that is they allow regulating the cognitive and affective learning activities by fostering the "thinking about thinking" and the "thinking about feeling" [HAC 99, BEL 10].

But meta-cognitive or regulative strategies are only developed later. As Sfard [SFA 91] highlights, procedural knowledge is, for most students, the first step. By referring to Piaget's study [PIA 72, p. 14], Sfard explains that the transition from procedural to declarative knowledge is a long and difficult process, accomplished through the assimilation of facts and procedures, and their consequent reification in explicit definitions. A further step is then needed to pass from declarative to meta-cognitive knowledge, since this latter is normally built upon repeated experiences of learning where the student has had the occasion to know him/herself, his/her learning strengths and weakness, as well as to identify valid patterns of solutions that can be transferred to isomorphic tasks, i.e. tasks being characterized by the same structure or number of steps [DAV 03], so that the student can transfer the acquired knowledge from a source domain to a target domain [MAY 02a]. Young students rarely possess meta-cognitive strategies in the early years of the school [KLA 00]. They thus need to be helped in acquiring these strategies, whose achievement make students closer to the ultimate objective of learning, which is learning to learn [HOS 08].

IBSE encourages the development of meta-cognitive strategies by inviting students to reflect on the activity, evaluate each phase of a project as well as the whole project, and perform near transfer (i.e. using the acquired knowledge and skills in slightly different contexts) and far transfer (i.e. using the acquired knowledge and skills in strongly different contexts) [WHI 98].

In our third study (section 3.3), the meta-cognitive dimensions is investigated by assessing pupils awareness of: (1) their understanding of the lesson content through the employed pedagogical tools, (2) their understanding of the pedagogical tool through the lesson content and (3) their ability to carry a new inquiry by replicating the steps followed in the first inquiry.

### 3.2.5. *Self-regulation*

Meta-cognitive or regulative strategies help to build self-regulation. In this sense, self-regulation is considered, within the frame of our research, a subdimension of meta-cognition. From an affective point of view, self-regulated learners approach tasks with confidence and resourcefulness. From a cognitive point of view, they proactively seek for knowledge and information that they feel are lacking. From a social point of view, they optimize their learning environment by adopting a cooperative behavior and seek advice. From a meta-cognitive point of view, they plan, set goals, self-monitor and self-evaluate. On the whole, they view learning as a systematic and controllable process and they accept greater responsibility for their achievement outcomes [ZIM 90]. Moreover, this kind of students tend adopt affective, social and meta-cognitive strategies that enhance their self-regulation in a self-oriented loop, which is oriented to reduce the gap between their goals and their outcomes [ZIM 89].

Therefore, in the last 20 years, unguided or minimally guided learning approaches encouraging pupils' self-regulation have become very popular and intuitively appealing and have started to replace instructional or guided approach. Advocates of minimal guidance or unguided approaches argue that "instructional guidance interferes with the natural processes by which learners draw on their unique, prior experience and learning styles to construct new, situated knowledge in line with their learning objectives" [ZIM 90].

However, a debate has been raised by research communities about the efficiency of this approach. Kirschner *et al.* [KIR 06] show that unguided approaches are less efficient than guided ones, due to empirically demonstrated reasons: without guidance, students have no point of reference to progress on their learning path, and can easily accumulate misconceptions. Moreover, the free exploration of a highly complex environment may generate a heavy working memory load that is detrimental to learning [KIR 06]. This is even more relevant for novice learners, who do not yet dispose of the mental schema that would help them to integrate new information.

A response to Kirschner and colleagues objections about constructivism comes from Hmelo-Silver *et al.* [HME 07] who state that not all the approaches categorized as minimally guided or unguided by Kirschner *et al.* are actually characterized by absence of guidance. For example, while discovery learning is mostly based on free exploration of a given subject, problem-based learning and inquiry-based learning employ scaffolding extensively thereby reducing the cognitive load.

Today, even if school programs oriented toward self-regulation have proved to be effective at primary school level (see [DIG 08] for a review), teachers still have difficulties in attributing much autonomy to young students, overall in crowded classes; in France, a primary class can contain up to 30 pupils, so that here learning autonomously in this case can generate chaotic and hardly unpredictable learning environments, where it would be difficult for a teacher to ensure that each pupil can properly access pedagogical tools, pose questions, express his/her opinion and gain a deep understanding of the subject at hand [GAU 12b].

However, 20 years of IBL projects have led to the consideration that self-regulation cannot be an all-or-nothing option in a teaching/learning approach: it should be gradually introduced into school curricula. This is to avoid unequal learning results like those observed by Sadeh and Zion [SAD 09]. These researchers followed a 2-year inquiry class focusing their investigation on conceptual change, intellectual flexibility and critical thinking. In their study, students were divided into two groups according to the kind of inquiry: guided inquiry and open inquiry. The data collected (*interviews, students' inquiry summary papers, logbooks and reflections*) revealed that open inquiry students showed significantly higher levels of performances in conceptual changes. However, the study's results indicated no significant differences in procedural understanding and in the affective dimension of learning.

As a solution, Rylander [RYL 12] proposes to clearly identify the differences between "cookbook style" – i.e. strongly guided learning, similar to following a receipt – and "inquiry experience" (Table 3.1) a self-regulated oriented program where levels of inquiry are progressively implemented (Table 3.2).

| Cookbook experiences | Inquiry experiences |
|---|---|
| Are driven with step by step instructions requiring minimum intellectual engagement of students thereby promoting rule-conforming behaviors. | Are driven by questions requiring ongoing intellectual engagement using high-order thinking skills, making for independent thought and action. |
| Commonly focus students' activities on verifying information previously communicated in class, thereby moving from abstract to concrete. | Focus students' activities on collecting and interpreting data to discover new concepts, principles or laws, thereby moving from concrete toward abstract. |
| Assume students will learn the nature of scientific inquiry by experience or implicitly; students execute imposed experimental designs that tell students which variables to hold constant, which to vary, which are independent and which are dependent. | Require students to create their own controlled experimental design; require students to independently identify, distinguish and control pertinent independent and dependent variables; promote students' understanding of the skills and nature of scientific inquiry. |
| Rarely allow students to confront and deal with error, uncertainty and misconceptions; do not allow students to experience blind alleys or dead ends. | Commonly allow for students to learn from their mistakes and missteps; provide time and opportunity for students to make and recover from mistakes. |
| Employ procedures that are inconsistent with the nature of scientific endeavor; show the work of science to be an unrealistic linear process. | Employ procedures that are much more consistent with authentic scientific practice: show the work of science to be recursive and self-correcting. |

**Table 3.1.** *Differences between strongly guided teaching/learning approaches and IBL learning approach according to Rylander [RYL 12]*

| Day | Type of activity | Example activity |
|---|---|---|
| Monday | Exploration experiences | Stations around the room that allow students to explore the concept of the lesson |
| Tuesday | Interactive demonstrations | Demonstrations and class discussion used to address misconceptions, familiarize students with new equipment, and model the experimental design |
| Wednesday | Discovery experiments | Student-designed experiment that seeks to discover a relationship that exists between two or more variables |
| Thursday<br>Friday | Application challenge | Challenge or context in which students must apply the relation that they have discovered |

**Table 3.2.** *Levels of inquiry in a weekly science curriculum according to Rylander [RYL 12]*

This is of course just one of the possible solutions to adopt for allowing students to enter straightforwardly in contact with science issues and thus become active learners who progressively take care of their own learning experiences.

However, self-regulation or autonomy does not mean that the pupil works alone, but rather that he/she seeks for appropriate guidance when needed, for example relying on different sources of supports to carry an inquiry, such as the advices of the teacher or a peer, the examples found in a manual or on the Web, etc. [FRI 13].

In our third study (section 3.3), self-regulation has been investigated by asking pupils if, during the making of the science project, they have recurred to: (1) teachers, (2) books or (3) peers to accomplish the different phases of the project, or (4) if they have worked on their own.

### 3.2.6. *RBI and inquiry-based learning*

Inquiry-based learning is hardwired into robotic-based activities: not only does RBI promote inquiries by increasing applications and connection of science to everyday life [ROB 05], but as Eguchi *et al.* [EGU 12, p. 372] remark, "students tend to encounter more opportunities for IBL based on real world problems with educational robots. This is because educational robots give students immediate, objective and unequivocal feedback as to whether their program and robot work or not. If the feedback from the performance of the robot indicates that the program is not working, it triggers their inquiry of why is not working".

Nevertheless, teachers cannot simply rely on the fact that inquiry automatically occurs during robotic activities. As Hmelo-Silver *et al.* [HME 07] state, robotic activities do not provide students with instant inquiry-based learning opportunities. Thus, scaffolding for structured activities is necessary. To this end, Williams *et al.* [WIL 07] have observed that even when giving students a general introduction to the process of scientific inquiry during a robotic camp, RBI has an impact on content knowledge, but not on scientific inquiry skills: at the end of the robotic camp, students showed increased physics knowledge, but when asked to carry an inquiry, they still proceed by intuitive solutions and by trials and errors instead of following the steps of the inquiry. The authors have given four different possible explanations for the results: (1) the novelty and attractiveness of robotic kits might have distracted students attention from the inquiry-related issues to the robotic-related issues; (2) the inquiry-related issues might have be introduced in a too general way due to lack of experience of teachers in IBSE; (3) acquisition of inquiry skills might be a long-term process that requires much more than a 2-week summer camp; (4) finally, the assessment instrument – based on quantitative and qualitative measures of robotics and physics knowledge, as well of problems solving abilities – could have been insufficient to test the acquisition of inquiry skills, since this instrument has been later criticized by its authors [KET 06] as not sensitive enough to detect students' learning of scientific inquiry. In order to reduce the novelty effect, Williams *et al.* [WIL 07] thus suggest allowing students an unstructured time to explore robots before starting the inquiry. Furthermore, they stress the necessity to foster the teaching of scientific inquiry by building long-term school programs based on IBSE.

Based on the study of Williams *et al.* [WIL 07], we suggest that the combination of RBI and IBSE in a long-term project should have a positive impact not only on domain-specific knowledge but also on fundamental science competencies. Moreover, we put forward that this combined frame can result to be beneficial for students learning processes in the cognitive, meta-cognitive, affective and social domains, because RBI and IBSE share high compatibility on three levels: pedagogical, curricular and structural.

#### 3.2.6.1. *Pedagogical compatibility: transversal competencies and learning domains*

According to a number of studies, educational robots can in principle give learners opportunities to improve their affective, cognitive, social and meta-cognitive abilities [DEN 94, DEN 01] without unnecessary cognitive loads [JUN 09].

Moreover, several authors have suggested that RBI enhances both students' subject-specific knowledge and transversal competencies. To this end, one of the most exhaustive models is one proposed by Denis *et al.* [DEN 01] who refer to the architecture of competencies by Leclerq [LEC 87] to show the potential of robotic-based activities on domain-specific and non-domain-specific competencies.

Taking in consideration these two points – i.e. the potential of RBI for the development of the four dimensions of learning and domain-specific and non-domain specific competencies (Table 3.3) – we have adapted the proposal of Denis *et al.* [DEN 01] by specifying, for each level of competencies and RBI objective, the concerned learning domain.

| Level of architecture of competencies | Learning domain | RBI objectives |
|---|---|---|
| Dynamic<br>Dynamic competencies are related to motivation, i.e. the pleasure a person experiences in doing things, in learning specific, demultiplicative or strategic competencies. This level is the most vulnerable: it can be easily affected. It is also the most "penetrating", i.e. the motor that drives the rest when facing a new domain which the learner has to enter. Those competencies correspond to the learner's initiative, will, pleasure and displeasure, perseverance, rigor, etc., including one's own image as a person being able and motivated to learn. | Affective | (Personal) project<br>Meaningful activities |
| Strategic<br>Strategic competencies are concerned with metacognition, i.e. knowing oneself (as a learner, as an actor, etc.), one's weaknesses and one's talents, and developing strategies to adapt to complex situations (for instance to choose which demultiplicative competence to use for learning in given circumstances). They concern planning (e.g. how much time will it take me to master a given subject, to do a specific work), problem solving (e.g. analysis and structuring the problem, decision making, etc.), communication and cooperation (how much and when do I need others? in which respect?) and self-estimation of knowledge (to know one's degree of expertise in a domain: what do I know? what do I not know?). | Meta-cognitive<br><br>Social | Problem solving and formalization of thought: structuring the problem, definition and test of hypotheses, conclusions.<br><br>Socialization: collaboration, sociocognitive conflicts, etc. |
| Demultiplicative<br>Demultiplicative competencies (learning tools) enable the learner to get information by him/herself and acquire more specific competencies: reading, listening, notes taking, communicating, interviewing, using the computer to consult a database or to produce texts, referring to guides, etc. | Cognitive | Reading, listening, notes taking, consultation of reference guides using the computer. |
| Specific<br>Specific competencies (elements of cognition skills) deal with specific contents (e.g. geography, history, physics and vocabulary of a language) and are hardly transferable. These specific competencies are infinite and a human being can (and has to) know only some of them. | Cognitive | Electronics components, types of electric circuits, programming instructions, production procedures, syntax of a given computer language. |

**Table 3.3.** *Levels of architecture of competencies [LEC 87] at play in RBI activities according to Denis and Hubert [DEN 01], with the corresponding four learning domains*

Additionally, with concern to self-regulation, Denis *et al.* [DEN 01] proposed to apply the regulation process by Leclerq [LEC 95]. This process includes six regulative phases at play in any learning/teaching process: (1) needs analysis or problem identification, (2) project definition, (3) planning, (4) action, (5) observation or measure and (6) amelioration. Since in RBI, the retroaction coming from the results of the evaluation will deal with one (or several) phase(s) of the regulation process, the authors have proposed to represent these phases in a loop rather than in a list (Figure 3.1).

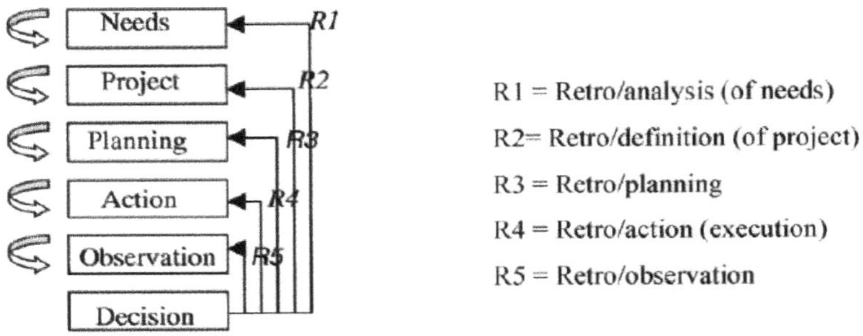


**Figure 3.1.** *Self-regulation loops in robotic projects according to Denis and Hubert ([DEN 01], adapted from [LEC 95])*

Denis *et al.* [DEN 01] also precise that the regulative processes can be accomplished by different means: discovering the material without guidance or with minimal guidance, imitating (e.g. *watching a video or observing a model and trying to reproduce it*) and referring to written guidelines or oral suggestions.

Another proposal of frame showing the deployment of the four learning domains within RBI comes from the Educational Robotic Applications [CAT 12], which stresses the importance of RBI for fostering students cognitive, social, personal and emotional skills in an authentic learning situation (Figure 3.2). Catlin and Blamires highlight that the ultimate goal of RBI, if its potential is fully deployed, is to enhance sustainable learning, that is learning in the longer term through the development of meta-cognition, life skills and learner self-knowledge. Moreover, as feedback is embedded in robotic activities (i.e. the fact that the constructed and programmed robot works or not is a feedback for the student who has carried the activity), RBI favors self-regulation [CAT12].

In this sense, empirical research has confirmed that RBI has a pivotal role for cognition and meta-cognition, as well as for affective and social aspects of learning. With respect to cognitive aspects, Eguchi *et al.* [EGU 12] have highlighted that robots allow pupils to externalize ideas or mental models to understand whether these ideas or models are wrong. In this sense, robotic kits like LEGO Mindstorms® and WeDo® allow pupils and students to build upon previous knowledge. Since these robotic kits are composed of robotic LEGO bricks that pupils have probably already encountered in their childhood, pupils and students deal with artifacts they are familiar with (the bricks) and progress toward something that they are new to (sensors and motors). Moreover, Slangen *et al.* [SLA 11] have observed that robotic-based activities allow to unveil pupils reasoning patterns because they have the capacity to provoke discourse and higher order thinking skills, such as analyzing, synthesizing, evaluating and performing causal reasoning. Meta-cognitive skills play an important role in RBI as well, since pupils and students are guided to replace the question "how do I solve this problem?" by the question "how do I make the robot solve this problem?" [MAT 04]. That is, by using the robot as a "transitional object" [PAP 80] or an "object to think with" [RES 96] learners tend to elaborate strategies of solution by taking a "cognitive distance" from the problem, and these strategies can thus be more easily generalized or transferred since the student use a third-person point of view rather than a first-person point of view, and this helps them to get a higher order perspective [GAU 16a]. Transfer is one of the most acclaimed skill in RBI [e.g. AHL 02], but its achievement occurs under specific conditions: activities must be organized by classes [ALI 09] and be highly structured [KHL88], so that students can learn to generalize solutions to a specific set of problems. Self-regulation is also considered pivotal to carry robotic activities: students learn to rely on different types of guidance – i.e. peer's guidance, information on different kinds of supports such as manuals, online documents and video [DEN 01] –

beyond the teacher's guidance. Furthermore, a number of works witness that RBI stimulates students interest [RI 09, PET 04] and encourage them to achieve their work – since, differently from an exercise book that students may close if they have difficulties to find solutions, a robot arouses a feeling of "I want it to work" [GAU 16a]. Additionally, robotic-based activities increase classroom interactions [CHE 03], are susceptible of reducing the gap between disadvantaged and non-disadvantaged students [MCD 12] and enhance students' attitudes toward science regardless of gender and ethnicity [JEW 12].

Overall, these theoretical and empirical studies show high pedagogical compatibility of RBI with IBSE in that both are supposed to enhance domain-specific and non-domain-specific knowledge and skills, and both aim at covering the learning process in its four main dimensions (cognitive, affective, social and meta-cognitive).

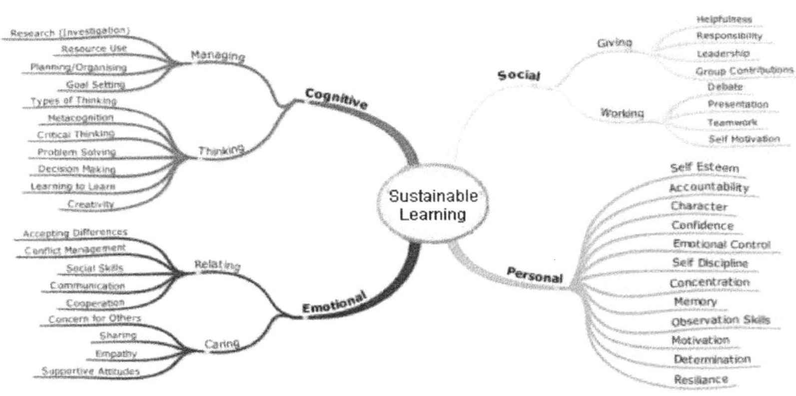


**Figure 3.2.** *Potential of RBI for sustainable learning, according to Catlin and Blamires [CAT 12]*

However, as discussed above (see Introduction and section 3.1) in order for IBSE to scaffold RBI at school, it is also necessary that they are compatible in terms of structure of the lesson and anchorage to school curriculum.

### 3.2.6.2. *Structural compatibility*

Among the RBI studies, we can find studies that offer an overview of the structure of the applied robotic-based activities, explained in terms of main processes or detailed in terms of phases. For instance, inspired by the observation of toddlers learning behaviors at kindergarten, Resnick [KIN 07] has identified a spiraling cycle of "imagine, create, play, share, reflect" as the main process describing an ongoing robotic activity. McDonald *et al.* [MCD 12] have employed this cycle to carry out robotic-based activities with Australian indigenous learners, adapting it to the local context (i.e. replacing "imagine" by "watch", since More and Williams [MOR 04] identified that the indigenous learners use to learn through observation and imitation rather than through verbal instruction).

Another example comes from Barker *et al.* [BAR 07a], who have borrowed from Kolb's [KOL 84] experiential learning theory to structure their robotics curriculum by five phases: (1) experience (i.e. do the activity), (2) share (i.e. *react and observe in a social context*), (3) process (i.e. *analyze and reflect upon what happened*), (4) generalize (i.e. *discover what was learned and connect to life*) and (5) apply what was learned to a similar or different situation.

Ribeiro *et al.* [RIB 08] also describe the main phases of theater robotic-based activities in which students use robots to represent traditional Portuguese tales. These phases were as follows: (1) preparation: children learn the basic concepts and functioning of robotic kits by building, programming and solving a number of exercises of increasing complexity; (2) dramatization of the play, organized into four main steps: programming of the characters; rebuilding of the robots for enhanced robustness; programming of the definitive version of the characters; integration of the wardrobe and final rehearsal; and (3) presentation of the project to the community and participation to science fairs or robotics festivals.

Finally, Kim and Baek [KIM 14] mention four main processes in robotic-based activities (planning, investigation, analysis and adjustments), while Alimisis [ALI 09] proposes a "praxeological organization" where students carry robotic activities that are grouped by "class of problems" and these classes include scientific and technological issues of growing difficulty.

Similarly, within the frame of the Pri-Sci-Net project[6] we have built RBI and IBSE robotic-based templates by dividing the lesson in three main phases: *engage*, *inquiry* and *evaluate* [RIE 16]. The *engage* phase includes pre-existing knowledge retrieval, investigation challenge and introduction to robotics. The *inquiry* phase includes the identification of a group challenge or inquiry, as well as the modeling, building, programming and testing of the robot. The *evaluation* phase involves a presentation of the group work, in which the students point out the main objectives of their project, what difficulties they encountered and how they overcame these difficulties; in this phase, students are also invited to set a debate, propose improvements, make use of the observed evidence and come back to the inquiry problem to find out which new knowledge and skills they have acquired with relation to it.

If we compare the structure of the robotic-based activities mentioned in the above-presented studies and projects that have the structure of typical IBSE activities – e.g. nine phases identified by Bells [BEL 10], or the iterated sequence "question–predict–experiment–model–apply" proposed by White and Frederiksen [WHI 98]; we can note a significant similarity in the main phases at play, as well in the idea that learning is not merely a linear result-oriented process but a cycle where improvements and new questions can be settled upon a first result [QUI 04, WIN 04, SCH 05], and where inquiries of growing complexity can be carried [ALI 09], thus giving students a concrete measure of their increasing knowledge and skills.

### 3.2.6.3. *Curriculum compatibility*

As Catlin and Blamires [CAT 12] state, if educational robots are to make a significant impact, they must be able to address two items: the curriculum and the assessment.

---

6 http://www.prisci.net/.

The combination of RBI and IBSE is a very recent path of research [RIB 12], thus we still do not dispose of rigorous evaluation methods to measure the effect of introducing this combined approach in French primary curricula.

In countries where scaffolding of RBI by IBSE has been applied, case studies showed that when coupled with IBSE, RBI is a valid pedagogical environment for both teaching and learning at school. That is, on the one hand, it helps teachers in adapting to the new inductive model of education which goes from facts to laws – thus employing concrete inquiry experiences to explain class of physical phenomena – rather than from laws to fact: as mobile technologies robots produces "facts" (movement of wheels, trajectories, etc.), and raise in this sense a more tangible understanding than software conceived to support students scientific inquiries, e.g. Web-Based Inquiry Science Environment [SLO 04], "Viten" [MOR 06], symphony [QUI 99] and Co-Lab [QUI 04, DEM 12]. On the other hand, RBI enhances students problem solving, mathematical and geometrical skills, and lets pupils familiarize themselves with computational thinking [DEM 12], which is becoming a fundamental skill, comparable with reading and writing: as an example, the new French Primary School Curriculum places computational language among the basic linguistic competencies of the first years, beside French language, foreign language, mathematical language and arts languages (see Appendix 2). Moreover, the RBI and IBSE combined approach has obtained profound effects on class learning dynamics; although this combined approach demands an effort in terms of logistics and renewal of a vision of education that needs time to root in schools, the final result is so gratifying for both teachers and learners that they agree it is worth the effort [EGU 12].

These impacts have mainly been measured through holistic qualitative measures or impressions of educators, such as observation of class behaviors and examination of "robotic" journals compiled by students [DEM 12, EGU 12]. Qualitative measures are of high interest in new domains of research where impacting factors have to be identified: they focus on processes, help to define key issues of investigation and analyze participants in their natural environments

[MER 98, RIB 08]. However, while assessment methods and corresponding quantitative and qualitative measures for IBSE start to be developed [MAR 04], there is not yet a rigorous evaluation of a combined RBI and IBSE approach.

In our research, we have opted for an assessment that takes into account the teacher's points of view as well as the pupils' points of view. Self-evaluation is an important part of assessment when investigating successful conditions of learning [MES 08]. Previous studies have shown that pupils and students are at ease when asked to evaluate subject-specific competencies (e.g. mathematics, physics and informatics) but experience difficulties when asked to evaluate high-order competencies related to the four learning domains: for example, they are not able to tell whether their planning and collaborative abilities are increased and they need guidance to express their point of view about this [SKL 03]. Therefore, in collaboration with primary teachers, we have designed an assessment questionnaire adapting the Pri-Sci-Net assessment methodology [GAT 16] to our specific research objectives and declining questions to allow pupils express their point of view on RBI and IBSE approach with respect to the four learning domains.

## 3.3. Study 3: impacts of ER on learning

### 3.3.1. *Aims and rationale*

The aim of this study is to evaluate the impact of RObeeZ, a primary school project combining IBSE approach and RBI environment, on primary pupils' learning processes and results. In particular, our objective was to assess the effects of RObeeZ on:

– *pupils subject-specific knowledge*: in order to do this, we have collected teachers quantitative evaluation of pupils' results on mathematics and science in the first and last trimester of the CM1 school year;

– *pupils transversal competencies*: in order to do this, we have collected teachers' qualitative evaluation on pupils' personal bulletins of competencies;

– pupils' development of the four dimensions of learning (cognitive, affective, social and meta-cognitive), through a specific questionnaire, based on the Pri-Sci-Net methodology assessment [MES 08], and co-built by the researcher and the teachers involved in this study.

The project is detailed below along with the hypotheses and the experimental design of the study.

### 3.3.2. *The RObeeZ project*

RObeeZ is a 2-year project combining ER as a technopedagogical environment and IBL as a teaching/learning approach.

Developed by a team of talented science and technology teachers of the primary school Louis Michel (Académie d'Amiens, Creil), the project has involved a class of 26 pupils, aged 9–10 years. The school belongs to the ÉCLAIR network (*Écoles, collèges et lycées pour l'ambition, l'innovation et la réussite*) which includes those suburban and rural schools, which beneficiates of a ministerial priority program to assure equal learning opportunities. The population of the Louis Michel school is thus partly constituted by children coming from disfavored social environments.

The choice of developing a project based on robotics and IBSE in this class was partly motivated by the wish of the teacher to create a stimulating learning environment for pupils who showed difficulties in problem solving, tended to apply mathematical procedures mechanically, and who during the previous year reported to believe that mathematics has no utility in real life. According to the teachers, these children also demonstrated limited interest in school subjects, could not express themselves properly and lacked awareness of the environment surrounding their town.

The general aim of RObeeZ was twofold: on the one hand, stimulating pupils to learn about the life of bees: their morphology, behavior and roles in a hive; on the other hand, introducing pupils to robotic technology by constructing and programming robotic kits.

Within this project, robots were thus intended as a tool to meaningfully convey science and technology contents and skills. The robots used were LEGO Mindstorms® and Robotami®. The programming software were LEGO NXT® and TronzCard®.

For their intrinsic features of constructible and programmable technologies, these modular robotic kits can be used to better understand the morphology and the behavior of a bee, as well as to acquire competencies in mechanisms and algorithms.

In this sense, the pedagogical objectives of RObeeZ have been defined by the teachers and the researcher involved in the project combining RBI and IBSE priorities (Table 3.4).

| RBI pedagogical objectives in RObeeZ | IBSE pedagogical objectives in RObeeZ |
|---|---|
| Develop children observation, analysis and design skills | Carry environmental observations |
| Raise the taste for problem solving | Raise the need for information search |
| Understand logic and mathematical operations | Apply scientific procedures (modeling, testing, etc.) |
| Learn to collaborate | Become aware of the impact of living beings on the environments and identify human civic responsibilities |
| Create, play and shape ideas | Reflect about present and future actions to preserve the ecosystem |

**Table 3.4.** *RObeeZ pedagogical objectives have been defined by combination of robotic-based instruction and inquiry-based science priorities*

Such objectives were operationalized into five main phases:

– *Preliminary phase*: In this phase, the teachers (1) retrieved the appropriate materials: robots, tablets and some exemplars of dead bees; (2) defined timetable: the project had a duration of three trimesters, and covered a major part of the CM1 (first year of the intermediary course at primary level) science and technology curriculum; (3) organized the class management: two teachers lead this project, with one teacher doing lessons about bees and guiding pupils in the construction of the robotic bees, and another one being in charge of guiding pupils in the programming of the robots.

– *Familiarization phase*: Pupils familiarized themselves with the functioning of the robots by constructing and programming simple models (e.g. vehicles) on the basis of the instructions contained in the LEGO Mindstorms® kit.

– *Modeling, construction and programming phase*: Pupils were assigned to two class sections: one section was in charge of constructing the robotic bees and the other one of programming them. The two sections were divided by groups of four pupils each. The groups of the first section was guided to search for information about bee morphology, and to make drawings of bees, at the beginning copying from books, and then doing it by themselves (Figure 3.3); these drawings were used as models to assemble bricks, sensors and motors (Figures 3.4(a)–(c)) to obtain prototypes of robotic bees (Figures 3.5(a)–(c)). The groups of the second section were guided to search for information about bee behaviors, and to schematize these behaviors and roles by natural languages and logic connectors (circles, arrows, etc.); these schemes were used as sequences for the programming scripts (Figure 3.6).

Five types of bees were considered: the queen bee, the mason bee, the foraging bee, the hive-guard bee and the nurse bee. According to their role, bees have specific behaviors and interactions, which the pupils had to transpose by means of scripts. The queen had to find the mason bee to start the building of an alveolus; the mason had to build or look for the alveolus (only if it was stimulated to do so by the queen); the foraging bee had to look for the pollen and to go back to the hive (only if it had found the pollen); the hive-guard bee had to let the foraging bee coming in (only of this latter had brought the pollen); the nurse bee had to receive the pollen collected by the foraging bee and to put it in the alveolus.

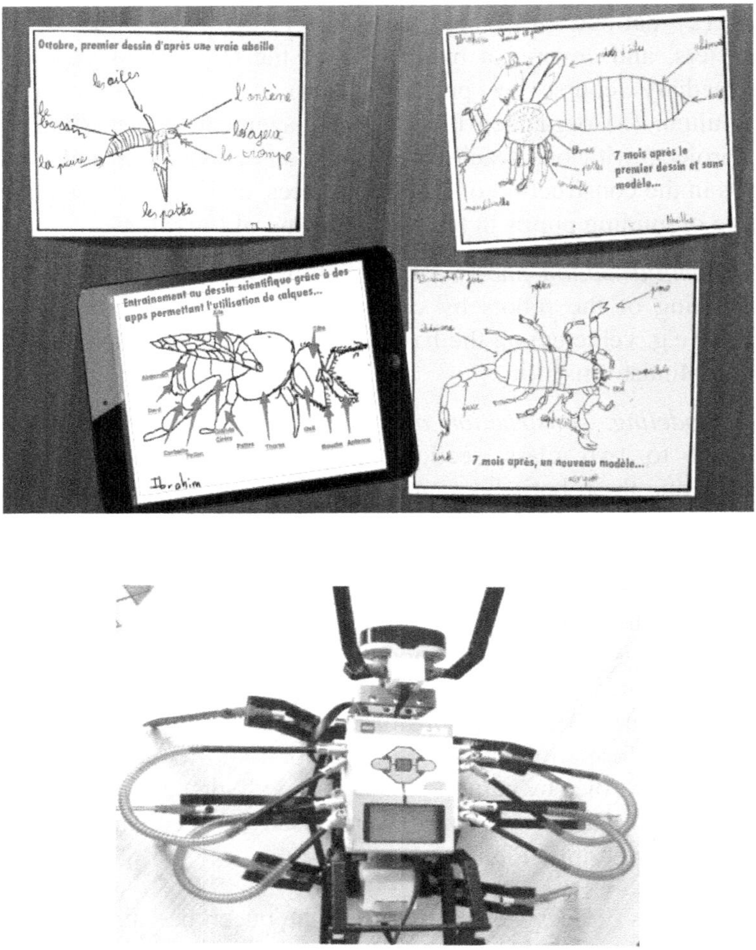


**Figure 3.3.** *Top: Drawings of bees by children in the modeling phase. In the first drawing (upper-left), students observed a real bee to draw a model; the second drawing (upper-right) was made by students without observing the bee, but simply recalling what they had seen 7 months before; the third drawing (lower-left) was realized by using an iPad app that allows to reproduce a carbon copy on a digital support; the fourth drawing (lower-right) was made after having built the first robotic bee prototype. Bottom: a first prototype of robotic bee made of Lego Mindstorms® components*

**Figure 3.4.** *The researcher and the students in the construction phase of the project*

Within each group, pupils played different roles, which were changed at each lesson in order to let each child apply all the targeted competencies: project leader, technician, programmer and materials supervisor. The project leader was in charge of coordinating the group, reading the instructions, formulating problems, recalling the delays, taking care of the workspace, welcoming the questions of the members, supporting the discussion and taking photos of the different stages of the project. Technicians constructed the robots, took notes and helped to correct eventual errors. Programmers were responsible for the tablets, programmed the robots and launched the programs. Materials supervisors had to select and position appropriate bricks, captors and motors for the technicians and to monitor the work, trying to anticipate the eventual technical needs:

– *Challenge phase*: Finally, the two sections meet in the so-called SHOWbeeZ (Figure 3.7) in order to combine their respective work: construction and programming. Here, the challenge was to make the five bees interact. Of course, several adjustments were necessary, and the challenge was considered accomplished when different types of bees managed to interact.

– *Exhibition, competitions and demonstrative lessons*: Groups also organized exhibitions to show their robotic prototypes, gave presentations during project competitions and were in charge of demonstrative lessons for college students.

RObeeZ was attributed the Parisian Prize of the Pedagogical innovation in April 2015[7].

**Figure 3.5.** *Further prototypes of robotic bees*

7 http://crdp.ac-amiens.fr/cddpoise/blog_mediatheque/?p=16430.

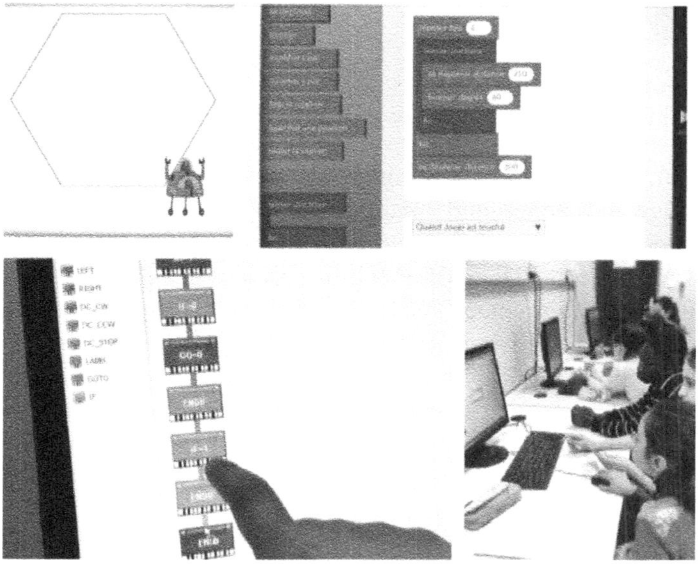

**Figure 3.6.** *Students coding the behavior of the robotic bees through TronzCard® interface*

**Figure 3.7.** *As a conclusive phase of the project, students of the construction section and students of the programming section combined their respective constructions and programs in the SHOWbeeZ*

### 3.3.3. *Hypotheses*

Upon consideration of the examined literature, the hypotheses of our third experimental study are as follows:

HYPOTHESIS 3.1.– Combining RBI and IBSE increases pupils knowledge in science and mathematics. As several researches have pointed out, RBI has great potential to enhance students understanding of STEM subjects [HUS 06, MIT 08, NUG 08, NUG 09, NUG 10, BAR 09] and life sciences [WIL 07, BRE 10]. This potential seems to be fully concretized when RBI is coupled with a suitable pedagogical approach [SUL 08], in particular with IBSE [DEM 12, EGU 12]. We thus predict that participating in RObeeZ would lead to an improvement of pupils' grades in the last trimester for mathematics (calculation, geometry, measures, problems and numbers) and science subjects.

HYPOTHESIS 3.2.– Combining RBI and IBSE enhances pupils competencies in science and mathematics, as well as their transversal competencies. Different studies witness that RBI can foster transversal competencies such as problem solving, team working, communication, planning, self-confidence and habits of mind [DEN 01, SKL 03, CHA 07, LIN 07, SUL 08, BAR 09]. However, while in RBI these transversal competencies are implicitly targeted, in IBSE they are explicitly pursued by putting the accent on the structure of the lesson and making students aware of this structure by providing them a model of the inquiry to follow [BEL 10]. This implies that, without IBSE, students do not easily acquire fundamental process skills [WIL 07] because they are not aware of the structure of the inquiry [HME 07]. Hence, we predict that pupils taking part in RObeeZ show an enhancement of competencies targeted in the last trimester: verbal and written understanding and expression, science and mathematical process skills, technological fluency, and autonomy and initiative taking. Furthermore, as previous research has pointed out [MCD 12], RBI contributes to reduce the gap between disadvantaged and non-disadvantaged students. Hence, we expect that RObeeZ improves in particular the learning results of those pupils who normally obtain low results.

HYPOTHESIS 3.3.– Combining RBI and IBSE favors the development of cognitive, affective, social and meta-cognitive dimensions of learning. Theoretical [DEN 01, CAT 12] and empirical studies [SLA 10, MCD 12, JEW 12] have pointed out that RBI involves learning in all its fundamental dimensions. In this sense, RBI and IBSE show high compatibility: inquiry-based science also underpins an integrated vision of learning [RIE 16] where students develops cognitive [HAR 00, SAD 09, MIN 10], affective [WIG 98, ALB 04] , social [DRI 94, DIL 99] and meta-cognitive skills [WHI 98, BEL 10], which include self-regulative skills [RYL 12]. For this reason, we predict that the combination of RBI and IBSE will determine high score attribution on the four learning dimensions questionnaire, filled by pupils at the end of the RObeeZ project. With respect to self-regulation, as previous studies witnessed that self-regulative processes in RBI take place through diversified forms of guidance rather than uniquely through teachers [DEN 01], we expect pupils to attribute high scores to questions concerning guidance by groups and different sources of documentation like manuals and online documents.

### 3.3.4. *Experimental design*

#### 3.3.4.1. *Participants*

25 pupils participated in the study, 19 boys and 6 girls. They belonged to the same primary class, from a suburban French school. Among these, seven showed strong learning difficulties, nine showed relative learning difficulties and nine showed no learning difficulties in the first trimester. They were all aged between 9 and 10 years.

#### 3.3.4.2. *Material*

Pedagogical material to carry out the project included robot and tablets: pupils used LEGO Mindstorms® and the NXT® interface to familiarize themselves with robot construction and programming. In the phase of design and realization of the robotic bees, they used Robotami® to build the robots and TronzCard® to control them. The robots were controlled through an iPad. During the project, pupils also used the school computers to gather information about bees and robots.

The assessment material included: (1) table of pupils' subjects-specific grades (mathematics: calculation, geometry, measures, problems, numbers; science) attributed by the teacher on a 0–20 scale; (2) bulletin of pupils' competencies progression, evaluated by the teacher as strong progression, or low progression (Table 3.5); (3) self-evaluation questionnaire – jointly designed by the researcher and teachers on the base of Pri-Sci-Net assessment methodology [MES 08] – in which pupils answered questions on the cognitive, affective, social and meta-cognitive domains of learning (Tables 3.6 and 3.7). The Pri-Sci-Net assessment methodology proposes an integrated assessment environment, where the researchers propose an evaluation tool (i.e. a questionnaire) based on the six IBSE Pri-Sci-Net principles (see section 3.2) and adapted to a specific RBI class project, while the teachers validate the questionnaire with respect to curriculum requirements, and pupils answer the questions at the end of the project thereby self-evaluating their learning experience. As pointed out in section 3.2.6.3, self-evaluation is a central ingredient of IBSE and RBI combined approach, since it facilitates pupils in developing awareness of high-order skills according to a holistic view of learning, which includes intellectual as well as emotional and relational aspects.

| Level 2 competence 1: Language | |
|---|---|
| Speaking | Be able to take part to a dialogue: speak in front of the class, listen to the others, express one's point of view |
| Reading | – Comprehend texts and instructions<br>– Identify explicit information in a text<br>– Infer implicit information in a text<br>– Search documents (books or multimedia) |
| Level 2 competence 3: Fundamental elements of mathematics and technology | |
| Organization and data management | – Read, interpret and build simple representation of data: tables, graphics<br>– Be able to organize mathematical and geometrical data to explain results<br>– Solve a problem |

| | |
|---|---|
| Carry a scientific and technological inquiry | Carry an inquiry: (1) observe, question; (2) manipulate and experiment; (3) formulate and test hypothesis; (4) argument and propose different solutions; (5) reason on the obtained results by using the appropriate scientific terminology in writing and discussing |
| Acquire and apply scientific knowledge in the appropriate contexts and activities | – Sciences<br>– Technological artifacts |
| Level 2 competence 4: Information and communication | |
| Employ informatics tools to work | Know and use the basic function of a computer and connected devices |
| Adopt a responsible attitude | Be aware of the issues and impacts of informatics tools and Internet and keeping a critical stance when using them |
| Create, produce, treat and use results | – Produce a digital document (text, image, sound)<br>– Use informatics tools to present a work |
| Be informed, look for documents | – Read a digital document<br>– Look for information by using the computer<br>– Be aware of resources and limits of Internet |
| Communicate, exchange | Employ ITC to communicate and exchange documents |
| Level 2 competence 6: Social and civic competencies | |
| Have a responsible behavior | – Respect class rules<br>– Respect the classmates |
| Level 2 competence 7: Autonomy and initiative | |
| Employ methods of work to be autonomous | – Respect simple instructions, working autonomously<br>– Be perseverant<br>– Try to self-evaluate one's work when carrying simple tasks<br>– Be capable of sustained attention (reading, music, spectacle, etc.) |
| Taking initiatives | Be engaged in individual and collective objectives |

**Table 3.5.** *Table of pupils' competencies progression, evaluated by the teacher at the end of the project. Extracted and translated from the official French "Personal bulletin of competencies" (see Appendix 5)*

<table>
<tr><td rowspan="2">Cognitive dimension</td><td>Ability to retrieve preexisting knowledge</td><td colspan="3">Ability to correct preexisting knowledge</td><td colspan="2">Ability to acquire new knowledge</td></tr>
<tr><td>Have you used knowledge you already had about bees or robots?</td><td>Did you correct some of your previous knowledge about bees?</td><td colspan="2">Did you correct some of your previous knowledge about robots?</td><td>Do you think you learnt something new about bees? If yes, what?*</td><td>Do you think you learnt something new about robots? If yes, what?*</td></tr>
<tr><td rowspan="2">Affective dimension</td><td colspan="4">Enjoyability</td><td colspan="2">Engagement</td></tr>
<tr><td colspan="4">Did you like the project?</td><td colspan="2">What kind of project would you like to do?*</td></tr>
<tr><td rowspan="2">Social dimension</td><td colspan="4">Communication</td><td colspan="2">Collaboration</td></tr>
<tr><td>During the construction phase, did you communicate with the pupils of your group?</td><td>During the construction phase, did you communicate with the pupils of the other groups?</td><td>During the programming phase, did you communicate with the pupils of your group?</td><td>During the programming phase, did you communicate with the pupils of the other groups?</td><td colspan="2">Do you think that your team did a good job?</td></tr>
<tr><td rowspan="2">Meta-cognitive dimension</td><td colspan="4">Relation between technology and pedagogical content</td><td>Autoevaluation</td><td>Transfer</td></tr>
<tr><td>Do you think that working on bees helped you to better understand the robot?</td><td colspan="3">Do you think that working with robots helped you better understand the bees?</td><td>Do you think that the class has succeeded the project?</td><td>Would be able to perform the different steps of the project to realize another project?</td></tr>
</table>

**Table 3.6.** *Self-evaluation questionnaire adopted in this study to assess the effect of the RObeeZ project on pupils' development of the cognitive, affective, social and meta-cognitive dimensions. Pupils were asked to attribute a score (from 0 to 5) as an answer to each question, with exception for questions indicated by an asterisk "*" symbol in the table: in this case, pupils had to provide open verbal answers*

| Self-regulation | |
|---|---|
| *Did you look for new information about bees:* | by yourself/with the teacher/with your group |
| *The main subject of the project was proposed by:* | the teachers/ you, the pupils |
| *Did you draw the model of the bee:* | by yourself/with the teacher/with the group |
| *Did you construct the robotic bee:* | by yourself/with the teacher/with the group/following the instructions on a manual or online documents |
| *Did you schematize the behavior of the bee:* | by yourself/with the teacher/with the group |
| *Did you program the bee:* | inventing your own code/using the code provided by the teacher/with the group/copying form a manual or on online documents/with your group |

**Table 3.7.** *The self-regulation questionnaire treated as a subsection of the meta-cognitive dimension section. Pupils were asked to attribute a score (from 0 to 5) to the options proposed in the right column*

#### 3.3.4.3. *Procedure*

Pupils started the RObeeZ project in the first trimester of the CM1 school year. The project activities covered two lessons of 2 h per week; sessions were intensified (three per week) during the SHOWbeez. After a first phase of familiarization, the class was divided into two sections: one section modeled and built five different types of robotic bees; another section programmed the scripts to control the behavior of the five robotic bees. Pupils' grades related to mathematics and science were collected by the teacher before the starting of the project and at the end of the project. Pupils transversal competencies were evaluated in the final phase of the project: the researcher extracted and translated relevant sheets from pupils personal bulletin of competencies (see Table 3.5) and asked the teacher to provide a qualitative evaluation of pupils' competence progression. Consequently, the teacher gathered the bulletins in two categories: strong competence progression – i.e. the bulletins of pupils who have shown considerable improvements of their transversal competencies in the last trimester – and low competence progression, i.e. the bulletins of pupils who have shown little improvements of their transversal competencies in the last trimester. The development of their cognitive, affective, social and meta-cognitive abilities have been observed through a self-evaluation questionnaire, distributed by the researcher and filled by students at the end of the project (see Tables 3.6 and 3.7). In order to avoid any bias in pupils' self-evaluation, these data were collected before their participation for the Prize for Pedagogical Innovation in April 2015. Informed consent was gained from the parents of pupils volunteering to participate in the study. Pupils were informed about the confidentiality and anonymity of their responses. All questionnaires sheets were stored securely and anonymously.

#### 3.3.4.4. *Data collection and analysis*

Three kinds of data were collected: (1) pupils results (grades from 0 to 20 – evaluation carried by the teacher) in the first and last trimester for the following subjects: mathematics (five grades: calculation, geometry, measures, problems and numbers) and science (one grade); (2) pupils' competence progression at the end of the project – two categories: strong progression and low progression (evaluation carried by the teacher, see Table 3.5); (3) the development of the four dimension of learning: cognitive, affective, social and meta-cognitive, with self-regulation considered as a subdimension of meta-cognition: scores from 0 to 5 (evaluation carried by the pupils, see Tables 3.6 and 3.7). This last measure was made through a multiple question survey, co-conceived by the teachers and the researcher on the basis of the Pri-Sci-Net assessment methodology.

## 3.4. Results

### 3.4.1. *Are the subject-specific knowledge and competencies of pupils enhanced at the end of RObeeZ project?*

To assess the impact of the RObeeZ project on pupils' grades in mathematics (calculation, geometry, measures, problems and numbers) and science, we have calculated the mean score in the first trimester ($M = 12.88$; SD = 4.59) and last semester ($M = 13.83$; SD = 3.64). The Shapiro–Wilk test reveals that the score distribution is not normal in the first trimester ($W = 0.96$; $P < 0.001$) as well as in the second trimester ($W = 0.96$; $P < 0.001$). Consequently, we have applied non-parametric test. The Wilcoxon test reveals a significant global increase in pupils scores in the last trimester ($W$ = 22490.5; $P < 0.001$).

When considering mathematics and science separately (Table 3.8), we observe a significant difference ($W$ = 2156; $P < 0.001$) between the first ($M$ = 12.16; SD = 5.43) and the last trimester ($M$ = 13.62; SD = 4.31) for mathematics grades. In particular, we register an increase in pupils for three (geometry, measures and problems) over five types of knowledge and competencies in mathematics; the two remaining types of knowledge and competence in mathematics for which we did not register improvements were calculation and numbers. While no significant difference ($W$ = 138.5; ns) can be remarked for science grades between the first ($M$ = 13.78; SD = 1.84) and last trimester ($M$ = 14.08; SD = 2.77). Hence, no significant improvements have been observed for science knowledge.

| Subject | First trimester | Last trimester | Wilcoxon test |
|---|---|---|---|
| Calculation | $M$ = 15.84; SD = 4.08 | $M$ = 13.73; SD = 3.57 | $W$ = 216.5; $P$ = 0.01 |
| Geometry | $M$ = 10.24; SD = 4.68 | $M$ =15.22; SD = 3.31 | $W$ = 0; $P$ = 1.3 |
| Measures | $M$ = 12.74; SD = 3.64 | $M$ = 14.44; SD = 2.88 | $W$ = 62.5; $P$ = 0.007 |
| Problems | $M$ = 7.08; SD = 5.04 | $M$ = 10.4; SD = 5.79 | $W$ = 23; $P$ = 0.00 |
| Numbers | $M$ = 14.88; SD = 4.69 | $M$ = 14.32; SD = 3.95 | $W$ = 183.5; $P$ = 0.34 |
| Science | $M$ = 13.78; SD = 1.84 | $M$ = 14.08; SD = 2.77 | $W$ = 138.5; ns |

**Table 3.8.** *Pupils results on mathematics and science subjects in the first and last trimester*

### 3.4.2. *Are pupils' transversal competencies enhanced at the end of the project?*

After RObeeZ, the teachers evaluated pupils' progression of the CM1 targeted competencies in language (speaking and reading), fundamental elements of mathematics and technology (organization and data management, carry a scientific and technological inquiry, acquire and apply scientific knowledge in the appropriate contexts and activities), information and communication (employ informatics tools to work; adopt a responsible attitude, create, produce, trait and use results; be informed, search for documents; communicate, exchange), social and civic competencies (respect class rules, respect the classmates), autonomy and initiative (employ methods of work to be autonomous, taking initiatives) (see Table 3.8 for a more detailed description of these competencies). Invited by researchers to provide a qualitative estimation of competence progression, the teacher divided their class in two groups of progression: strong progression, 56% of the pupils (i.e. 14 pupils); low progression, 44% of the pupils (i.e. 11 pupils). Among the 56% showing strong progression, 20% (i.e. five pupils) were considered by the teacher as having strong difficulties in the first trimester, 20% (i.e. four pupils) were considered having medium difficulties in the first trimester, and the remaining 16% (i.e. four pupils) had no difficulties in the first trimester. Among the 44% showing low progression in the last trimester, 12% (i.e. three pupils) were considered having strong difficulties in the first trimester, 4% (i.e. one pupil) was considered having medium difficulties in the first trimester, and 32% (i.e. eight pupils) were considered having no difficulty in the first trimester.

Overall, strong competence progression was mostly registered for pupils with difficulties in the first trimester. It thus seems that the combination of RBI and IBSE was particularly beneficial for these pupils, while those who obtained good learning results in terms of transversal competencies before the project showed less improvements in this sense.

### 3.4.3. *Does the combination of RBI and IBSE have an impact on the four dimensions of learning?*

The Shapiro–Wilk test reveals that the scores distribution for the cognitive ($W = 0.963$; ns), social ($W = 0.965$; ns), meta-cognitive ($W = 0.972$; ns) dimensions, and self-regulative subdimension ($W = 0.975$; ns) is normal. Therefore, we have applied a parametric test for these dimensions. The scores distribution for the affective dimension is not normal ($W = 0.445$; $P < 0.001$), so we have carried out non-parametric test for this dimension. As we can see in Figure 3.8, the impact of the project on learning dimensions is particularly significant for the affective dimension, followed by the social, cognitive and meta-cognitive dimension.

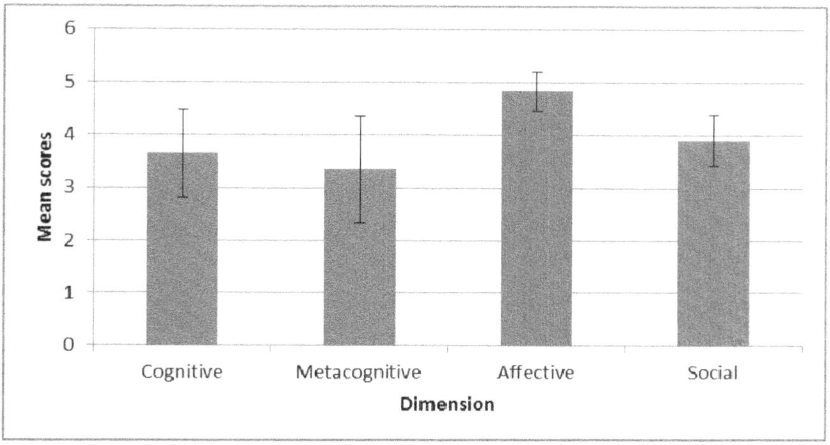


**Figure 3.8.** *The impact of the RObeeZ project on the cognitive meta-cognitive, affective and social dimension according to pupils' self-evaluation as retrieved from our questionnaire*

A more detailed analysis (means and standard deviation for each question composing the four dimensions) is shown in Tables 3.9 and 3.10.

The aspect of the cognitive dimension that has obtained the highest score is "ability to correct preexisting knowledge about robots" ($M$ = 4.12; SD = 1.12), followed by "ability to correct preexisting knowledge about bees" ($M$ = 3.48; SD = 1.26) and "ability to retrieve pre-existing knowledge about robots and bees" ($M$ = 3.36; SD = 0.99). It thus seems that the project gave pupils the opportunity to build new knowledge upon previous knowledge, reorganizing these latter according to new scientific schema, with relation both to robotics and life science aspects of the project. When asked what they had learnt about bees, 32% (i.e. eight pupils) answered that they learnt the five roles of the five types of studied bees; 20% (i.e. five pupils) said that they had learnt specific behaviors of a specific bee (e.g. only the queen bee, or only the foraging bee); 20% (i.e. five pupils) pupils said that they had learnt the morphological structure of the bee; 12% (i.e. three pupils) said that they had better understood the bee's reproduction; 8% (i.e. two pupils) reported they had understood the function of pollen and flowers; 4% (i.e. one pupil) said he learnt how bees move; finally, 4% (i.e. one pupil) wrote that once he had observed how bees live and he had created a robotic bee he was no longer scared of bees. These results show that pupils acquired specific knowledge about bees, probably related to the class-section they belonged to: pupils in the construction section acquired knowledge about morphology, while pupils in the programming section acquired knowledge about the behaviors of bees.

| | | | | | |
|---|---|---|---|---|---|
| Cognitive dimension | *Ability to retrieve pre-existing knowledge* | *Ability to correct pre-existing knowledge* | | *Ability to acquire new knowledge* | |
| | Have you used knowledge you already had about bees and robots? | Did you correct some of your previous knowledge about bees? | Did you correct some of your previous knowledge about robots? | Do you think you learnt something new about bees? If yes, what? | Do you think you learnt something new about robots? If yes, what? |
| Results | *M* = 3.36; SD = 0.99 | *M* = 3.48; SD = 1.26 | *M* = 4.12 SD = 1.12 | * | * |
| Affective dimension | *Enjoyability* | | | *Engagement* | |
| | Did you like the project? | | | What kind of project would you like to do? | |
| Results | *M* = 4.84; SD = 0.37 | | | * | |
| Social dimension | *Communication* | | | | *Collaboration* |
| | During the construction phase, did you communicate with the pupils of your group? | During the construction phase, did you communicate with the pupils of the other groups? | During the programming phase, did you communicate with the pupils of your group? | During the programming phase, did you communicate with the pupils of the other groups? | Do you think that your team did a good job? |
| Results | *M* = 4.2; SD = 1.25 | *M* = 2.28; SD = 1.46 | *M* = 4.32 SD = 1.02 | *M* = 2.52; SD = 1.56 | *M* = 4.68; SD = 0.64 |
| Meta-cognitive dimension | *Relation between technology and pedagogical content* | | *Achievement evaluation* | *Transfer* | |
| | Do you think that working on bees helped you to better understand the robot? | Do you think that working with robots helped you better understand the bees? | Do you think that the class has succeeded the project? | Would be able to perform the different steps of the project to realize another project? | |
| Results | *M* = 3.56; SD = 1.26 | *M* = 3; SD = 1.22 | *M* = 4.48; SD = 1.15 | *M* = 3.48; SD = 1.22 | |

**Table 3.9.** *Scores on the four learning domains: pupils' self-evaluation*

| Self-regulation | | | | |
|---|---|---|---|---|
| *Did you look for new information about bees:* | by yourself | with the teacher | with your group | |
| Results | $M = 3.36$; SD = 1.25 | $M = 3.68$; SD = 0.9 | $M = 3.32$; SD = 1.37 | |
| *The main subject of the project was proposed by:* | the teachers | you, the pupils | | |
| Results | $M = 4.76$; SD = 0.8 | $M = 1.36$; SD = 1.03 | | |
| *Did you draw the model of the bee:* | by yourself | with the teacher | with the group | |
| Results | $M = 2.2$; SD = 1 | $M = 3.84$; SD = 1.02 | $M = 2.2$; SD = 1 97 | |
| *Did you construct the robotic bee:* | by yourself | with the teacher | with the group | following the instructions on a manual |
| Results | $M = 2$; SD = 1.68 | $M = 2.36$; SD = 1.7 | $M = 3.84$; SD = 1.62 | $M = 1.8$; SD = 1.68 |
| *Did you schematize the behavior of the bee:* | by yourself | with the teacher | with the group | |
| Results | $M = 2.72$; SD = 1.64 | $M = 2.68$; SD = 1.62 | $M = 3.24$; SD = 1 61 | |
| *Did you program the bee:* | inventing your own code | using the code provided by the teacher | copying form a manual | with your group |
| Results | $M = 1.72$; SD = 1.45 | $M = 3.88$; SD = 1.4 | $M = 3.48$; SD = 1.73 | $M = 3.4$; SD = 1.75 |

**Table 3.10.** *Scores on self-regulation, a subdimension of the meta-cognitive domain: pupils' self-evaluation*

Affective dimensions obtained high scores ($M = 4.84$; SD = 0.37): the class reported to have extensively appreciated the project. Furthermore, when asked whether they would have liked to carry out further robotics projects, all the pupils gave affirmative answers. In particular, 20% (i.e. five pupils) wished to carry projects about animal-like robots (ants, horses, felines and butterflies); 20% (i.e. five pupils) wished to carry projects about human-like robots; four students said they would have liked to continue the RObeeZ project; 12% (i.e. three pupils) proposed projects about fast vehicles (cars and motorcycles); 4% (i.e. one pupil) proposed an ubiquitous robotics project (i.e. sensors detecting what happens in an environment); 4% (i.e. one pupil) wanted to realize a football robotic project; finally, 4%

(i.e. one pupil) wished to make a movie on robot. These results on affective dimension show that pupils enjoyed the project and felt engaged, thus wishing either to continue the RObeeZ project or to carry out another project on a different theme.

The aspect of social dimension that has obtained the highest score is collaboration ("Did you think your group has done a good job?" $M = 4.68$; SD = 0.75), followed by communication within the group ("Did you talk with the pupils of your group during the programming phase?" $M = 4.32$; SD = 1.03; "Did you talk with the pupils of your group during the construction phase?" $M = 4.20$; SD = 1.26). The aspect of social dimension which has obtained the lowest score is communication between the groups ("Did you talk with the pupils of the other groups during the programming phase?" $M = 2.52$; SD = 1.56) and ("Did you talk with the pupils of the other groups during the construction phase?" $M = 2.28$; SD = 1.46). These scores witness that pupils perceived a strong sense of cooperation and achievement. Furthermore, they focused on task-related communication, that is they exchanged ideas and opinions only with pupils of their own group.

The aspect of meta-cognition which has obtained the highest score is *achievement evaluation* ("Do you think that the class has succeeded the project?": $M = 4.48$; SD = 1.15), followed by *relation between pedagogical content and technology* ("Do you think that working on bees helped you to better understand the robot?": $M = 3.56$; SD = 1.26), *transfer* ("Would be able to perform the different steps of the project to realize another project?": $M = 3.48$; SD = 1.22) and again *relation between technology and pedagogical content* ("Do you think that working with robots helped you better understand the bees?" $M = 3$; SD = 1.22). These results point out that pupils were extremely satisfied with the general achievement of the project, which implied non-obvious coordination of two sections of the class, the construction and programming ones. Furthermore, they were confident that they could transfer the acquired competencies to new projects. Finally, while the observation and modeling of bees was perceived by pupils as useful to better understand the functioning of the robotic device, the benefit of working on robots to better understand the morphology and behavior of bees was perceived as more limited to some extent.

Concerning self-regulation, pupils searched for information about bees and robots mostly with the teacher (*M* = 3.68; SD = 0.9); information search alone (*M* = 3.36; SD = 1.25) and in group (*M* = 3.32; SD = 1.37) obtained slightly lower scores. Pupils also agreed on the fact that the main subject of the project was proposed by the teacher (*M* = 4.76; SD = 0.8), rather than by the class (*M* = 1.36; SD = 1.03).When modeling, they did it mostly following the suggestions of the teacher (*M* = 3.84; SD = 1.02), and, to a minor extent, with their group (*M* = 2.2; SD = 1.97) or alone (*M* = 2.2; SD = 1). When constructing, they did it mostly with the group (*M* = 3.84; SD = 1.62), and to a minor extent by following the suggestions of the teacher (*M* = 2.36; SD = 1.7), alone (*M* = 2; SD = 1.68) and following the instructions on a manual (*M* = 1.8; SD = 1.68). When schematizing the behavior of the bee, they did it mostly with the group (M = 3.24; SD = 1.61) and to a minor extent alone (M = 2.72; SD = 1.64) and with the teacher (*M* = 2.68; SD = 1.62). When programming the bees, they did so mostly using examples of code provided by the teacher (*M* = 3.88; SD = 1.4), or copying from manuals and online documents (*M* = 3.48; SD = 1.73), or with their group (*M* = 3.4; SD = 1.75); pupils rarely invented their own code (*M* = 1.72; SD = 1.45).

These results show that in the RObeeZ project a part of the activities were mainly teacher led, while another part was mostly pupils led. Teacher-led activities were as follows: searching for information, deciding the main subject of the project, modeling and programming. Pupil-led activities were as follows: constructing and schematizing the behavior of the bee. Other sources of guidance, like manuals and online documents, were considered useful in the programming phase, but not in the construction phase. The work group was also extensively applied in constructing, schematizing and programming. Overall, pupils rarely worked alone: they were supported either by their teachers or by their peers and manuals, and these forms of guidance often coexisted.

Finally, an analysis of the correlations among the four learning dimensions revealed that no correlation exists among these dimensions (Table 3.11), and this confirmed that these dimensions were distinct items of analysis.

| Correlations among the four dimensions of learning | | |
|---|---|---|
| Cognitive/meta-cognitive | $r^2 = 0.148$; | $P = 0.48$ |
| Cognitive/affective | rho = 0.198; | $P = 0.34$ |
| Cognitive/social | $r^2 = 0.175$; | $P = 0.40$ |
| Meta-cognitive/affective | rho = 0.053; | $P = 0.80$ |

**Table 3.11.** *Analysis of correlation among the cognitive, affective, social and metacognitive dimension of learning*

## 3.5. Discussion

This study aimed to assess the impact of RObeeZ, a 1-year project combining RBI and IBSE on learning in a primary class. Previous studies demonstrated that, despite RBI having a promising potential in enhancing students subject-specific [RES 91, HUS 06, WHI 07, WIL 07, MIT 08, RIB 08, RUS 08, ALI 09, BAR 09, BRE 10, NUG 10] as well as transversal knowledge and competence [DEN 01, SKL 03, CHA 07, LIN 07, RIB 09] to develop cognitive, affective, social and meta-cognitive dimensions of learning [DEN 01, CHE 03, PET 04, SLA 10, CAT 12, RIE 16], this potential stays unexpressed until RBI is supported by a suitable pedagogical approach [WIL 07, SUL 08, BEN 12]. Among the different educational approaches in current trends of education (e.g. project based, problem based, discovery based and hands-on), IBSE seems to be the most appropriate to scaffold robotic-based activities: not only has this approach proven to foster learning, since it holistically includes a number of principles that are separately emphasized by other approaches, but it also shares pedagogical, curricular and structural compatibility with RBI to a great extent.

Therefore, in our third experimental study, we hypothesized that the combination of RBI and IBSE in a long-term class project would have resulted in significant enhancement of knowledge and competencies, as well as the high-order skills composing the four dimensions of learning.

With respect to mathematics knowledge and competencies, results show that the RObeeZ project had a significant impact on pupils scores in geometry, measures and problems, but not on calculation and numbers. Furthermore, no significant improvements have been observed for science scores. Hence, our first hypothesis was partially confirmed. This first result is in line with previous studies on impacts of an RBI and IBSE combined approach on mathematical skills [DEM 12] and seems to suggest that while inquiry-based robotics activities are useful to increase pupils mastering of complex mathematical issues, like arithmetical as well as geometrical problems and measuring, they are less useful to train pupils to do more simple tasks such as calculations and estimation of quantities. This might be explained by the fact that these tasks are often automatized when working on control interfaces and computer software, so that pupils do not need to carry them manually. In this sense, IBSE robotic activities allow pupils to focus on stimulating tasks, while more instrumental and uninteresting tasks are achieved through software tools. On the other hand, the lack of improvement in the science grade score is not in line with studies assessing the impact of robotics on life science knowledge and competencies [WHI 07, BRE 10]. Two possible explanations of this result can be considered. First, while in the pupils' school bulletin mathematics knowledge and competencies are evaluated through a composite grade (including numbers, calculation, geometry, measures, problems), science is assessed by a single grade; this might have prevented to take into account the multiple knowledge and competencies included in science learning. Second, this project covers only one of the life science topics of the CM1 curriculum; other topics (e.g. light and shadow, energy and chemical solutions) have not been taught in this class by the RBI and IBSE method. The science grade thus concerns not only pupils learning results not only with relation to the RobeeZ project but also with relation to learning performances on other topics that have been taught by a more traditional approach, so that it was not evident for pupils how to transfer the newly RBI- and IBSE-acquired skills to these topics. Therefore, it would be interesting, in further research, to assess whether pupils' science grades improve when the RBI and IBSE approach is applied to a larger part of the science curriculum.

With respect to transversal knowledge and competencies, results show significant improvements in language, fundamental elements of mathematics and technology, information and communication, social and civic competencies, autonomy and initiative. These improvements mainly concern pupils with strong and medium difficulties, whereas teacher's evaluation of proficient pupils stayed unchanged. This result is in line with the literature on the importance of RBI and IBSE to provide equal learning opportunities and reducing the gap between disadvantaged and non-disadvantaged students [MCD 12], as well as between different ethnicities [JEW 12], as it was the case for the RObeeZ class. However, the fact that no particular improvement was registered for proficient students might open to questions about the potential of RBI and IBSE combined approach for this kind of students. An answer to this question has emerged during the postproject interviews with the two teachers being in charge of the project. To this concern, teachers reported that the project triggered a number of important changes in the attitudes of the pupils, whether they were proficient or not, leading to profound transformation of their way of learning, and that these changes are not easily evaluable through traditional score attributions. The changes that teachers observed were described as strong curiosity toward new subjects, improved ability to express their ideas, disposition to care for their peers and engagement in the project, during the lesson and beyond (many pupils continued to test the robotic prototypes and programs at home, without any demand from the teacher and wished to continue the project even after its official deadline). Furthermore, the teachers themselves felt highly engaged throughout all the project: although they initially had to face the fact that pupils were much more at ease in using robots than they were themselves, the teaching/learning environment was so stimulating that new insights for teaching were continuously autogenerating by the new class dynamics; furthermore, the activities carried in the project seemed to fulfill those theoretical principles of pedagogy that every teacher seeks to apply, but that are in fact the most difficult to apply (e.g. logistics and need to stick to curriculum). According to these teachers, the combination of RBI and IBSE helped in overcoming these obstacles since it fostered pedagogical innovation while taking into account the curriculum demands, and because it endorsed a proper equilibrium between guided and autonomous learning that prompted teachers to jointly work toward logistic solutions with regard to class management (e.g. dividing the class in two classes, so to ensure a rational organization of the activities), and that stimulated children to take responsibility of learning materials and group dynamics. This qualitative evaluation provided by teachers in the interview witnesses that the coupling of RBI and IBSE may engender an interesting renewal of future learning environments. However, it also points out the need for new methods of assessment that enable teachers and researchers to identify the crucial factors of this transformation and recognize good practices. These methods should take in account, beyond subject-related knowledge and competencies, also important attitudinal aspects of learning, as those ones mentioned by the teachers: curiosity, disposition to care for the other, ability to formulate and share ideas, engagement, constancy, etc.

With respect to the four dimensions of learning, results show that, from a cognitive point of view, the project had a positive impact on pupils' capability to retrieve and correct pre-existing knowledge about robots and, to a minor extent, about bees. This is in line with studies pointing out that, as tangible devices supporting empirical IBSE activities, robots allow to externalize mental models [PAP 80] and to, if necessary, acknowledge that these models are wrong and must thus be corrected, however frustrating this could be for a pupil [EGU 12]. Furthermore, when acquiring new knowledge, this knowledge was rather specific to the task attributed to their own class section: pupils in the construction section did not learn to program, and pupils in the programming section did not learn to construct. This organization of the class in two sections has undoubtedly facilitated the class management for the two teachers in charge of the project. Nonetheless, to give all the pupils the opportunity to learn both mechanics and programming, it would be important to switch the sections of the class in a further project.

From an affective point of view, just like other studies have demonstrated (e.g. [PET 04]) pupils expressed high enjoyment and engagement in the project: they liked to take part, and they expressed the wish to continue the RObeeZ project, or to carry out new projects on human-like, animal-like or vehicle-like robots. This can be considered as a non-trivial result: because RObeeZ is a long-term project, the excitement that can generally be adduced to the novelty effect of robotic tools [WIL 07]; in this case, it is confirmed as a long-lasting appreciation through a longitudinal study.

From a social point of view, pupils gave a positive evaluation of their collaborative group and reported to communicate extensively with pupils of their group, as has been observed in other studies [DEN 01, CHE 03] but not without the pupils of the other groups. Again, this might witness that pupils were focusing on their group task, so they felt the need to communicate their ideas and solution to make their group progress; furthermore, the fact of having precise roles (project leader, technician, programmer and responsible of the material) could have made their communication more structured.

From a meta-cognitive point of view, pupils showed satisfaction for the project achievements, and they felt capable of transferring the acquired competencies to start another project. This is in line with studies showing that transfer of competencies is possible on projects of the same type and level of complexity [ALI 09]. They also reported that the study of bees helped them to better understand the functioning of robots, and, to a minor extent, that the use of robots has helped them to better understand the life of bees. However, since many of them also reported to have discovered unknown things about bees (see results about the cognitive dimension in section 3.4.3) their answers can be explained by the fact that they considered that there were more unknown issues to understand about bees than about robots, the former being something that they have already encountered in their life, and that could be thus perceived as less difficult to understand than robots.

Furthermore, concerning self-regulation, results show that the guidance of the teacher was stronger in specific activities like information searching, definition of the main subject of the project, modeling and programming. Pupils were more autonomous (i.e. they relied on manuals, online documents, or on the advices of the members of their own group) for activities like constructing and schematizing the behavior of the bee. As Denis *et al.* [DEN 01] have stated, robotic activities naturally lead to use multiple supports for progressing in the project, so this result confirms our third hypothesis.

Overall, the results about the impact of the RObeeZ project on the four learning dimensions are in line with previous literature [DEN 01, CAT 12, RIE 16]. However, this impact has been evaluated on the basis of pupils' subjective impressions. For an objective evaluation, further research might assess in this class the concrete ability of pupils to transfer the acquired knowledge, competencies and high-order skills in a new project for the CM2 curriculum of the incoming year.

# Conclusion and Perspectives

This book aimed to investigate the emerging field of ER from a psychological perspective. Three main learning paradigms of ER have been examined: *learning robotics*, *with robotics* and *by robotics*. Due to the novelty of this field, fundamental issues of research have been approached:

– what is an educational robot: its ontological and pedagogical status, according to students' representation;

– to what extent do users accept the robot as a functional and social agent and which factors are susceptible of influencing the dynamics of human–robot trust;

– how do educational robots impact on learning processes (cognitive, affective, social and meta-cognitive dimensions of learning) and results (knowledge and competencies).

Results of the three experimental studies that we carried out to answer to these research questions show that:

– *Learning robotics* fosters a more nuanced judgment on the ontological status of a robot, but a more definite judgment about its educational status: after having constructed and programmed a robot, students' assignment of robots to the category of living entities is more graded, while attribution of pedagogical roles (object, tool and companion) to robots is more clear-cut.

– *Learning with robotics* raises users trust in the robot as functional agent, while trust in the robot as a social agent is rather limited. Desire for control, attitude toward social influence of robots, and type of interaction scenario did not have an impact on trust.

– *Learning by robotics* enhances mathematics knowledge and competencies, has a strong impact on the affective, social, cognitive and meta-cognitive dimensions of learning, and triggers a profound transformation of students' attitudes toward learning and teachers' attitudes toward teaching. However, these effects are not the results of robotic-based activities alone, but of the scaffolding by an appropriated pedagogical approach.

The research contribution of this research work can be discussed on three levels: (1) research methodology, (2) future educational applications and (3) human–robot interaction design.

## C.1. The necessity of a cross-disciplinary methodology

Our three experimental studies are based on cognitive and social psychology as well as on educational methods. This choice has been determined by the cross-disciplinary nature of ER, and also by the necessity of treating fundamental issues in such a new-birth discipline. In fact, as recent studies have pointed out [DIN 15b], robotics-related research involves a variety of methodological techniques according to the specific issue at play. For example, in his experiments aimed at investigating whether human beings could consider a robot as a friend or a colleague, Dinet [DIN 15b], employs attitudinal questionnaires as well as spontaneous verbalizations, video classification and physiological responses (i.e. Galvanic skin response). Throughout his study, the author highlights the necessity of identifying an interdisciplinary frame of research to cover the different robotic-related research issues. Philosophy, psychology and ethics would be applied to better understand emotions and interactions with robots; sociology, anthropology and design would be useful to treat the issue of robots' acceptance; ergonomics could provide directions to study robots' usability; computational science, cybernetics and artificial intelligence would be useful for determining the utility of the robots. The complementarity of all these methods is indispensible for a deeper understanding of as complex a research subject as educational and social robotics.

Our point of view is that, within ER, the analysis of usages, interactions and impacts of robots for learning should rely on methods that fit the specific educational paradigm at hand. For the paradigm *learning robotics*, what is relevant is not merely to what extent the robotic kits facilitate the acquisition of notions and abilities in engineering but which perceptions and beliefs are underpinned by the current usages of robots, and how individuals could contribute to the enhancement of these technologies in order to become intentional learners rather than passive users [BER 00]. To this aim, traditional methodological techniques from cognitive psychology (e.g. categorization tasks and word/image association) could be combined with other techniques normally used to examine the functioning of creative thought – e.g. the EPoc battery of tests, conceived to engage children in exploratory divergent activities and integrative convergent activities – [LUB 10] – but also with techniques from Educational Ergonomics that are meant to evaluate users' capabilities of improving their own educational technologies (e.g. [SMI 07]). In the paradigm *learning with robotics* the nature itself of humanoid and animal-like robots as entities at the boundary between living and non-living requires a joint analysis of explicit or subjective information (e.g. verbal answers) and of implicit or objective information (e.g. automatic or unconscious behaviors) retrieved respectively from questionnaires and behavioral observations. Such a joint analysis, that has already obtained interesting results on the inconsistency between explicit judgment and implicit behaviors in the interaction between humans and computers displaying social cues [NAS 93, NAS 97, NAS 05], might be extremely useful for revealing presuppositions of the interaction between humans and robots, whose fuzzy ontological status encourages understanding of the relationship between overt and unconscious behaviors of users. Finally, in the paradigm *learning by robotics*, Developmental Psychology methods[1] are required to identify learners needs as well as the real and lasting impacts of new technologies on cognitive development, as well on the meta-cognitive and emotional and social stages of the learners [PAS 13]; Educational Science methods also play an important role to set up school programs that could reduce the digital divide and to elaborate assessment tools that are coherent with the 21st century educational requirements in terms of societal and working skills [VEN 00, PAR 01]; furthermore, educational ergonomics is pivotal for the design and evaluation of interfaces that allow a customizable learning program [MCL 10], as recent initiatives are increasingly putting forward, to allow students to develop what they are good at and remediate what they are not good at[2].

## C.2. Broadening the field of robotics' educative applications

We have proposed that robotics kits employed in the educational paradigm *learning by robotics* are a promising tool for tuition, but their potential is actually realized only when robotic-based activities are scaffolded by an appropriate pedagogical frame, such as IBSE. Our third study (RObeeZ project Pri-Sci-Net, section 3.3) has been conceived and realized before the ongoing curriculum reform, which sets strongly inquiry-oriented programs and which encourages the use of digital technologies as teaching and learning tools starting from 2016. As we have defended throughout this book, this new vision of education will need to be supported by appropriate educational materials and practices. The use of robots should be accompanied by software that are specific to the school subject, so to adapt the didactical content to this new technology. For example, future developments of the RObeeZ project may include the exploitation of software expressly conceived for primary and secondary school to analyze animal's complex behavior through video captures and

1 For instance, longitudinal observation of the learners' behavior – e.g. verbalization during problem solving, attitudes toward their peers, etc. – accompanied by qualitative analysis of key moments of the learning process like failure or achievements by examining the assimilation and accommodation phases [PIA 67].

2 http://www.bbc.com/news/technology-34151045.

categorizing tools (*Animal Landlord*®), [GOL 02]: students produce fine-grained segmentation of the animal behaviors in each video, and compare and contrast their analyses. The main idea underpinned by the *Animal Landlord*® software is to scaffold the task of behavior analysis by making the investigation model explicit. Combining this software with LEGO Mindstorms® and WeDo® may result in a powerful technopedagogical environment that allows merging the affective and social potential of robotic kits-based activities with the cognitive and meta-cognitive support provided by this kind of interface.

Furthermore, although mostly applied to scientific tuition, an inquiry approach can also be employed for humanistic subjects [BEA 01]. For example, American high schools have introduced inquiry to study literature: the class chooses one or more novels and completes group projects and seminars around these novels by questioning the peers and the teachers on the psychology of the personages, the narrative dynamics, the author style and reflecting about contemporary cultural issues such as cohabitation of different ethnicities[3]. The opening of inquiry-based learning to subjects other than science finds parallels in ER: as Rusk and colleagues point out in their call for "new pathways into robotics" [RUS 08], robotic activities can be extended to humanistic and artistic tuition. In this sense, some examples of this kind of educative application start to emerge: for example, Ribeiro *et al.* [RIB 08] propose a robotic theater where robots are constructed and programmed to reproduce traditional Portuguese tails. Similarly, Hromada and Gaudiello [HRO 14] have proposed a story-teller robot for kindergarten. These proposals can be ascribed to a broader field of application in educational technologies called technology-mediated narrative environments for learning [DET 06], which in the past have relied mainly on software as narrative tools, and which now include robots as tools. In conclusion, applications of educational robots are still in their infancy. This brings us again to the Resnick *et al.* [RES 96] list called "Twenty things to do with a robotic brick" (see Introduction) that these authors have conceived as a contemporary answer to the older list of Papert and

3 https://www.teachingchannel.org/videos/inquiry-based-teaching-with-literature.

Solomon [PAP 72] called "Twenty things to do with a computer". This list included a variety of activities such as constructing and programming a haunted house, robotic collars for dogs and systems to water the plants. At the end of the list, the 20th activity says: "Think about other 20 activities to do with a robotic brick". This invitation, addressed to students as well as to educators and researchers, seems to ironically refer the recursive functioning of a computer program, and therefore to warn parents and decision makers: if the use of educational technologies at school is today a source of apprehension with respect to consequences on pupils' and students' abilities in terms of reasoning, attention, concentration and socialization [TUR 11], the 20th point of the Resnick *et al.* list suggests that these technologies can on the contrary make students' and pupils' learning experience extremely imaginative and engaging.

## C.3. New perspectives for human–robot interaction design

Our Study 1 (section 1.3) and 2 (section 2.3) have showed that, within the paradigms *learning robotics* and *learning with robotics*, a robot needs to have a precise roles in order for learners to consider it as a useful technology at school and to interact with it as a companion. In this sense, the idea of an artificial companion that is functionally and socially adapted to today's educational contexts can be taken into account only if this robot has a specific role, acknowledged by the class, for example that of carrying out exact measures to test a given hypothesis in a science project. Additionally, instructing the robot for a specific role can be itself a pedagogical activity for students, through a "learning by teaching" pedagogy, just like ongoing projects of research show. For instance, the RASPO project (*Robot pour l'Accompagnement Scolaire PerSonnalisé* [Robots for Personalized Guided Learning]) is based on interaction with a Nao robot to solve arithmetical problems through everyday situation tasks based on numbers, bills and coins (see [RUG XX]). Another interesting initiative comes from CS Unplugged – computer science without computers:[4] this Web site is a collection of free learning activities that employ games, puzzles, crayons and physical exercises (e.g. forming a

4 http://csunplugged.org/.

line of pupils that represent a computer string) to teach concepts such as binary numbers, algorithms and data compression, separated from the distractions and technical details of having to use computers. A socially interactive robot might be an ideal companion for this kind of activities, because it would be assigned a precise role (e.g. representing the number 0 or 1 in a binary string), and because pupils might find it amusing to explain what an algorithm is to a robot. Finally, *learning with robots* is becoming a relevant educational paradigm in research about remediation of cognitive and affective handicap [BOU 14].

However, at the present stage of robotics development, it is difficult to foresee whether these robots, and thus this specific type of interaction, will be concretely applied at school. Not only might costs related issues concerning humanoids and animal-like robots determine a preference for more affordable technologies like robotic kits, but beyond, the idea of interacting with a robot whose interactive abilities are rather limited does not necessarily raise the enthusiasm of educators and decision makers: despite the robot triggering an initial curiosity among learners because of its human-like or animal-like aspect, this initial curiosity would be transformed into deception due to lack of reactivity of the robot, and so learners' interest would decrease [FRI 14]. This makes a point, above all if we consider that the research advancements brought by educative robotics laboratories have engendered new prototypes of robots (e.g. Cublets[5], TERN[6] and Yana & Bo), that are cheaper and easier to use in terms of control and interaction. These new robots are hardly classifiable within the three educational paradigms treated in the present work, since they are characterized by interactions that are not comparable to those with humans, neither to those with animals, nor to those with other technological devices, but they are still susceptible of being perceived as having an intentional behavior (e.g. [LEV 15, LEV XX]). For example, the Cublets – that have been conceived to facilitate the understanding of complex systems dynamics as a result of simple systems dynamics – can be "programmed by construction": by assembling one block with another (movement block, sound block,

5 http://www.modrobotics.com/cubelets/.
6 http://hci.cs.tufts.edu/tern/.

light block, etc.), it is thus possible to control a robot that can take different aspects and behaviors. This highly intuitive mode of control, namely *tangible programming* [HOR 11, SUL 15], is bringing a new "fourth-generation technology", after the bricks kits, the mechanics bricks and the constructible and programmable kits. Tangible programming robots cannot be used to learn mechanics, electronics and programming, neither to socially interact with the robot as a classmate nor to acquire knowledge and competencies in traditional school subjects. But they can be used, at the present stage of their development, to realize mobile interactive sculptures. Or we can imagine further exploitations, such as teaching chemistry by creating tangible representations of chemical reactions. Such an approach to ER is a top-down approach: to use these robots we do not have to deal with bottom-up construction and programming (e.g. assembling wheels and motors, creating algorithms), as instead it happens with robotics kits, but to combine ready-to-be-used blocks simply regulating their features (intensity of light and sound, etc.).

Here, history seems to repeat: from "black-box" to "transparent" to black-box technologies again. But looking closer to the present trends of technologies beyond robotics (e.g. ubiquitous connected technology, that is technology integrated in our environments, like advanced devices in smart houses, controllable by personal technology, for example by our smartphones), we may remark that history is not exactly repeating: instead of simply getting back to black-box, we are establishing a different kind of transparency. Such transparency is no longer meant as the access to the internal structure of the robot through the computer, but as the extent to which the technology allows a symbiotic relation with the user and his/her environment [BRA 10].

In this sense, we might expect that new educational technologies will be symbiotic with the classroom environment, inherently interactive, adapted to students' learning profiles, extensively customizable, capable of connecting students to one another, as well as to the world outside school (for example imagine a robotic modular desk, where students could lead projects through software of group work management, by scheduling tasks and deadlines, retrieving online information, programming devices like sensors for measurements or taking video and audio-registration, and connecting to the robotic desks of other schools engaged in similar projects).

Interactivity, customizability and connectivity are of course not an end in themselves, but rather powerful instrumental features that can support a less "schoolish" and more *situated learning* [AND 96, MCL 06, SAD 09] by valorizing intellectual, emotional and collaborative strengths of students and replacing passive and cumulative process of knowledge and competence acquisition.

Finally, we believe that a full exploitation of these three paradigms of ER is possible only if a systemic change takes place, that is if the technical development of new educational tools is accompanied by a critical revision of educational objectives, ways of meaningfully integrating technologies in the lesson, and assessment priorities. These priorities, as pointed out in our third experimental study, should target not only knowledge and competence but also students' and teachers' life-long attitudes on the basis of the new educational needs in contemporary society.

# Appendices

# Appendix 1

## Twenty things to do with a programmable brick [RES 96]

1) Create a "haunted house". Attach a programmable brick to the door to make creaking sounds whenever the door is opened. Program another brick to drop spiders on people when they walk through the door. Build a LEGO platform for a pumpkin, and program a brick to drive the pumpkin around the room.

2) Connect sensors to various parts of your body. Then, program a programmable brick to monitor your heartbeat and breathing as you walk and run. Or program the brick to play different sounds when you move different parts of your body.

3) Take a programmable brick with you to measure the pH level of the water in local streams, or the noise levels at a local construction site.

4) Create a LEGO musical instrument. The instrument might have buttons like a flute, or a sliding part like a trombone, or a completely new interface that you invent. Start by writing a simple program so that the programmable brick plays different notes (or melodies) when you move different parts of the instrument. Then, enhance the program so that the brick improvises on your notes. Or program the brick to play "rounds" (by playing a second copy of your notes with a delay).

5) Put a programmable brick and light sensor on the door to keep track of the number of people that enter the room. Then program the brick to greet people as they enter the room (with music or digitized speech).

6) Set up a weather station on the roof of the building.

7) Use a programmable brick to find out if the light really does go off when you shut the refrigerator door.

8) Attach a programmable brick to an ashtray, and program it to play a coughing sound whenever anyone uses the ashtray.

9) Build a remote-controlled LEGO car. Use a standard television remote control to communicate (via infrared transmission) with a programmable brick in the car.

10) Create an "intelligent room" that automatically turns on the lights when someone walks in the room. (Here's one approach. Build a LEGO machine that turns on the light switch, and connect it to a programmable brick. Use another programmable brick to detect when anyone enters the room. Use infrared transmission to communicate between the two bricks.)

11) Use a programmable brick to control a video camera (via infrared transmission). Program the brick to make a time-lapse video of a plant growing (taking a few frames every hour or day).

12) Use a programmable brick to program your VCR.

13) Send secret messages across the room to someone else who also has a programmable brick.

14) Put a brick on your dog's collar and collect data about your dog's behavior. How much time does your dog spend running around? Discuss whether experimenting on your dog is ethical.

15) Use a brick to record your dog barking. Then, put the brick in a remote-controlled LEGO car. Play the barking sound when the LEGO car gets near a cat. How does the cat react?

16) Build a LEGO creature that you can interact with. Program the creature to act in different ways when you clap once, or clap twice, or shine a light in its "eyes".

17) Build a LEGO creature that explores its environment. Program the creature to find the part of the room with the most light or the highest temperature. Next, put a plant on your LEGO creature, so that the plant will always move to the part of the room with the most light (or the highest temperature). Use other sensors to monitor the growth of the plant.

18) Build a LEGO machine that can water your plants, and then program a brick to make the machine water the plants every few days.

19) Create a game where each player carries a programmable brick. Program the bricks so that they give instructions to the players, and send messages from one player to another.

Think up 20 more things to do with a programmable brick.

# Appendix 2

## The French primary curriculum reform

Primary school curricula in France have been based, until 2015, on the "Socle commun des connaissances et compéteneces", a set of knowledge and competencies that have to be acquired in CP (preparatory course, 6 years old), CE1, CE2 (elementary course, 7–8 years old), and CM1 and CM2 (intermediary course, 9–11 years old). This set of knowledge and competencies has been established upon the indications of OECD-PISA (Organisation for Economic Cooperation and Development-Programme for International Student Assessment[1]) and is based on seven areas: (1) mastering the French language, (2) practicing a foreign language, (3) knowing principal elements of mathematics and scientific cultures, (4) applying information and communication techniques, (5) acquiring a humanistic culture, (6) be aware of social and civic norms and (7) demonstrate autonomy and initiative.

Assessment in French primary school is traditionally by written and oral exercises. Knowledge acquisition for each school subject is measured by attributing grades (from 1 to 20) on a personal

1 http://www.oecd.org/fr/edu/scolaire/programmeinternationalpourlesuividesacquisde selevespisa/.

notes schedule; competence acquisition is evaluated through qualitative judgment on a personal bulletin of competences.

However, as the 2013 PISA report witnesses, performance of French students have dropped to an average position in international mathematics tests with a score down 16 points from 2003; France now ranks 25th in the index. Moreover, there is a real need to enhance the educational results of immigrants and disadvantaged students. Additionally, French students seem to be globally unprepared for 20th-Century skills because teachers have difficulties in meaningfully integrating technologies in daily lessons[2].

Therefore, a number of projects have been outlined to improve the quality of education in French schools and to prepare students for new 21st Century professional profiles: for example, the EU project *Pollen*[3] and the national project *La Main à la pâte*[4] with the latter being initiated by the French Academy of Sciences under the guidance of the Nobel Prize winner Georges Charpak. These projects promote the reforming of science teaching in primary schools based on the inquiry approach. Pri-Sci-Net belongs to this category of projects and shares their vision of education, but, unlike them, it introduces robotics as a tool for inquiry-based lessons.

Furthermore, in order to meet the new international standards of education, such as the Next Generation Science Standards[5] a renewal of the "*Socle commun des connaissances et compétences*" will take place starting from 2016. This renewal makes the primary curriculum significantly more inquiry oriented: students will be guided to learn how to carry out a scientific inquiry through collaborative group projects around a main theme, which will be defined in the form of problem solving and which will require developing abilities of observation, data collection, hypothesis-making testing, reasoning

2 http://www.oecd.org/pisa/keyfindings/pisa-2012-results.htm.
3 http://cordis.europa.eu/result/rcn/51592_en.html.
4 http://www.fondation-lamap.org/.
5 http://www.nextgenscience.org/next-generation-science-standards.

and communicating; moreover, students will learn not only to use technologies, but to create technologies, and to employ them as tools for subject-related and transversal learning. Official programs and new assessment instruments will be released only in 2016, but a summary of the incoming reform can be found on the Ministerial Platform Eduscol[6].

6 http://eduscol.education.fr/cid86943/nouveau-socle-commun-pour-2016.html.

# Appendix 3

## Pre-questionnaire filled by students before taking part in the 3-day robotics event RoboParty® (section 3.3)

**Name/School:** **Sex: F/M**

**Age:** **Nationality:**

**1) How often have you watched or read something about robots?**
(Never) 0 1 2 3 4 (Often)

**2) How often do you play with robots?**
(Never) 0 1 2 3 4 (Often)

**3) Are the robots you played with similar to one or more of the following?** (Check the right box(es))

□  □  □  □ 

**4) How often have you built or programmed a robot?**

(Never) 0 1 2 3 4 (Often)

**5) How many times in the past have you attended robot-related lessons, exhibitions, competitions or other events?**

(Never) 0 1 2 3 4 (Often)

**6) These images make you think of a robot**

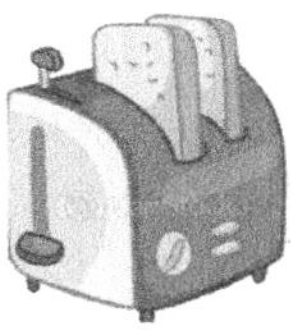

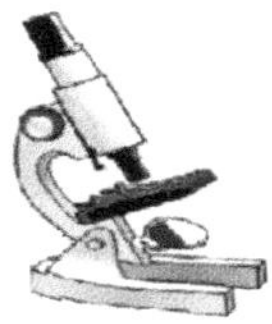
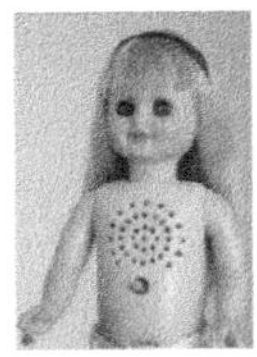

(Few) 0 1 2 3 4 (A lot) (Few) 0 1 2 3 4 (A lot) (Few) 0 1 2 3 4 (A lot) (Few) 0 1 2 3 4 (A lot)

(Few) 0 1 2 3 4 (A lot) (Few) 0 1 2 3 4 (A lot) (Few) 0 1 2 3 4 (A lot) (Few) 0 1 2 3 4 (A lot)

**7) A robot is:**

(A) (Less advanced) 0 1 2 3 4 (More advanced) than a machine

(B) (Less advanced) 0 1 2 3 4 (More advanced) than a computer

(C) (Less advanced) 0 1 2 3 4 (More advanced) than a scientific tool

(D) (Less advanced) 0 1 2 3 4 (More advanced) than a plant

(E) (Less advanced) 0 1 2 3 4 (More advanced) than an animal

(F) (Less advanced) 0 1 2 3 4 (More advanced) than a human

**8) Which name would you give to the robot? (Check the right box(es))**

□ A person name □ An animal name □ An invented name

**9) Do you think that a robot can**

(A) Do what you ask it to do (No) 0 1 2 3 4 (Yes)

(B) Take decisions (No) 0 1 2 3 4 (Yes)

(C) Do only what I told him to do (No) 0 1 2 3 4 (Yes)

(D) Switch on by itself (No) 0 1 2 3 4 (Yes)

(E) Make mistakes (No) 0 1 2 3 4 (Yes)

(F) Grow up (No) 0 1 2 3 4 (Yes)

**10) In classroom, you would like to use robots**

(No) 0 1 2 3 4 (Yes)

**11) You would prefer:**

(A) Using the robot as a tool (No) 0 1 2 3 4 (Yes)

(B) Learning with the robot as a classmate (No) 0 1 2 3 4 (Yes)

(C) Creating and programming a robot (No) 0 1 2 3 4 (Yes)

**12) Building and programming a robot can improve your achievements in:**

(A) Mathematics (No) 0 1 2 3 4 (Yes)

(B) Literature (No) 0 1 2 3 4 (Yes)

(C) Science (No) 0 1 2 3 4 (Yes)

(D) Technologies (No) 0 1 2 3 4 (Yes)

(E) History (No) 0 1 2 3 4 (Yes)

(F) Geography (No) 0 1 2 3 4 (Yes)

(G) Art (No) 0 1 2 3 4 (Yes)

(H) Religion (No) 0 1 2 3 4 (Yes)

(I) Music (No) 0 1 2 3 4 (Yes)

**13) Robotics activities in the classroom could help you in:**

| | |
|---|---|
| (A) Becoming aware of what I am able to do | (No) 0 1 2 3 4 (Yes) |
| (B) Working in a group | (No) 0 1 2 3 4 (Yes) |
| (C) Being more attentive while learning | (No) 0 1 2 3 4 (Yes) |
| (D) Expressing what I think | (No) 0 1 2 3 4 (Yes) |
| (E) Planning | (No) 0 1 2 3 4 (Yes) |
| (F) Solving problems | (No) 0 1 2 3 4 (Yes) |
| (G) Being creative | (No) 0 1 2 3 4 (Yes) |

**14) What will you do with the Bot'n Roll robot after RoboParty?**

## Post-questionnaire filled by students after having participated in the 3-day robotics event RoboParty® (section 3.3)

**1) Did you enjoy RoboParty?** (No) 0 1 2 3 4 5 (Yes)

**2) How much time did you spent in building & programming the robot during each of the 3 days?**

| | | | |
|---|---|---|---|
| 1st day: | □ <4 h | □ Between 4 and 7 h | □ >7 h |
| 2nd day: | □ <4 h | □ Between 4 and 7 h | □ >7 h |
| 3rd day: | □ <4 h | □ Between 4 and 7 h | □ >7 h |

**3) These images make you think of a robot**

(Few) 0 1 2 3 4 (A lot) (Few) 0 1 2 3 4 (A lot) (Few) 0 1 2 3 4 (A lot) (Few) 0 1 2 3 4 (A lot)

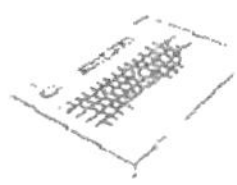

(Few) 0 1 2 3 4 (A lot) (Few) 0 1 2 3 4 (A lot) (Few) 0 1 2 3 4 (A lot) (Few) 0 1 2 3 4 (A lot)

**4) Do you think that a robot can**

(A) Switch on by itself (No) 0 1 2 3 4 (Yes)

(B) Do what you ask it to do (No) 0 1 2 3 4 (Yes)

(C) Grow up (No) 0 1 2 3 4 (Yes)

(D) Make mistakes (No) 0 1 2 3 4 (Yes)

(E) Do only what I told him to do (No) 0 1 2 3 4 (Yes)

(F) Take decisions (No) 0 1 2 3 4 (Yes)

**5) In classroom, you would like to use robots** (No) 0 1 2 3 4 (Yes)

**6) Would you prefer:**

(A) Learning with the robot as a classmate (No) 0 1 2 3 4 (Yes)
(B) Using the robot as a tool (No) 0 1 2 3 4 (Yes)
(C) Creating and programming a robot (No) 0 1 2 3 4 (Yes)

**7) Which name would you give to the robot? (Check the right box(es))**

☐ A person name ☐ An animal name ☐ An invented name

**8) Building and programming a robot can improve your achievements in:**

(A) Mathematics (No) 0 1 2 3 4 (Yes)

(B) Art (No) 0 1 2 3 4 (Yes)

(C) Literature (No) 0 1 2 3 4 (Yes)

(D) History (No) 0 1 2 3 4 (Yes)

(E) Science (No) 0 1 2 3 4 (Yes)

| | |
|---|---|
| (F) Geography | (No) 0 1 2 3 4 (Yes) |
| (G) Religion | (No) 0 1 2 3 4 (Yes) |
| (H) Music | (No) 0 1 2 3 4 (Yes) |
| (I) Technologies | (No) 0 1 2 3 4 (Yes) |

**9) Robotics activities in the classroom could help you in:**

| | |
|---|---|
| (A) Becoming aware of what I am able to do | (No) 0 1 2 3 4 (Yes) |
| (B) Working in group | (No) 0 1 2 3 4 (Yes) |
| (C) Being creative | (No) 0 1 2 3 4 (Yes) |
| (D) Being more attentive while learning | (No) 0 1 2 3 4 (Yes) |
| (E) Expressing what I think | (No) 0 1 2 3 4 (Yes) |
| (F) Solving problems | (No) 0 1 2 3 4 (Yes) |

**10) A robot is:**

| | | | | | | |
|---|---|---|---|---|---|---|
| (A) (Less advanced) | 0 | 1 | 2 | 3 | 4 | (More advanced) than a machine |
| (B) (Less advanced) | 0 | 1 | 2 | 3 | 4 | (More advanced) than a computer |
| (C) (Less advanced) | 0 | 1 | 2 | 3 | 4 | (More advanced) than a scientific tool |
| (D) (Less advanced) | 0 | 1 | 2 | 3 | 4 | (More advanced) than a plant |
| (E) (Less advanced) | 0 | 1 | 2 | 3 | 4 | (More advanced) than an animal |
| (F) (Less advanced) | 0 | 1 | 2 | 3 | 4 | (More advanced) than a human |

# Appendix 4

The WoZ GUI was organized in several tabs, each dedicated to a specific task, such as controlling the robot movements (gaze, hands movements, posture, its speech, its face expressions, etc.). The GUI events are elaborated by the actionServer module and others developed by the authors in previous works [IVA 14a, IVA 14b].

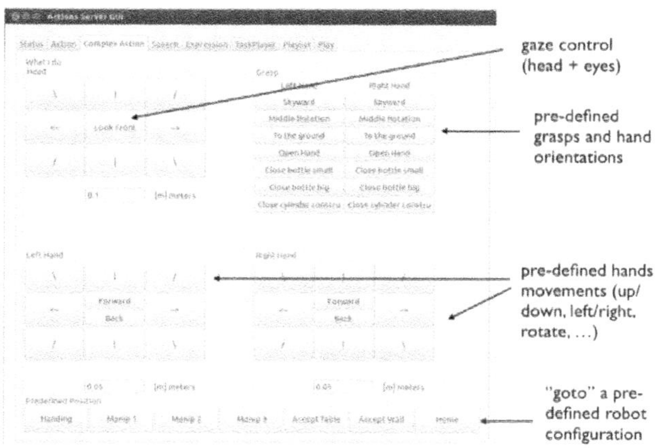


**Figure A4.1.** *WoZ GUI: the tab dedicated to the quick control of gaze, grasps and hands movements in the Cartesian space. The buttons send predefined commands to the actionsServer module [IVA 14b]. The buttons of the bottom row allow the operator to bring the robot in predefined postures (whole-body joint configurations): they were preprogrammed so as to simplify the control of the iCub during the experiments, in case the operator had to "bring it back" to a predefined configuration that could simplify the interaction for the subjects. They were useful also for prototyping and testing of the experiments*

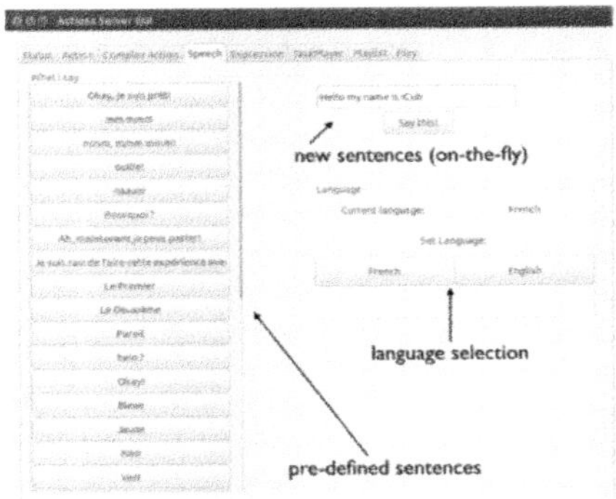


**Figure A4.2.** *WoZ GUI: the tab related to the robot's speech. The operator can choose between a list of predefined sentences and expressions, or he/she can type a new sentence on-the-fly: this is done to be able to quickly formulate an answer to an unexpected request of the participant. The operator can switch between French and English speech (at the moment, the only two supported languages), even if in the experiments of this paper of course the robot was always speaking French*

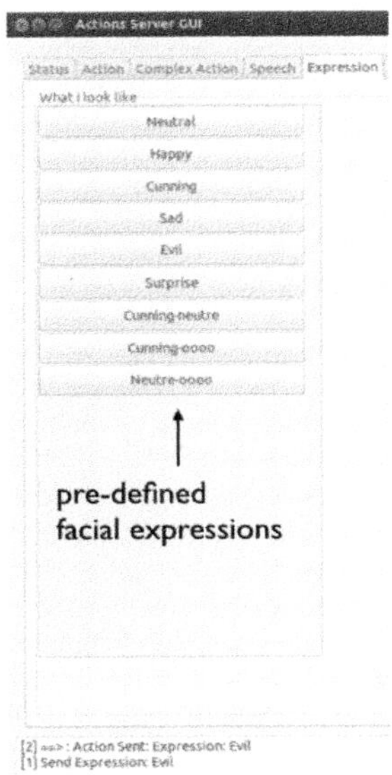


**Figure A4.3.** *WoZ GUI: the tab related to facial expressions. The list of facial expression along with their specific realization on the iCub face (the combination of the activation of the LEDs in eyelids and mouth) is loaded from a configuration file*

## The desire for control questionnaire (study 2, section 2.3)

The order of the questions follows the original questionnaire proposed by Burger and Cooper [BUR 79]. The left column reports the original questions in English. In the below table, the right column reports our translation of the questions in French.

| No. | Questionnaire item in English | Questionnaire item in French |
|---|---|---|
| 1 | I prefer a job where I have a lot of control over what I do and when I do it. | Je préfère un travail où j'ai un contrôle important sur ce que je fais et quand je le fais. |
| 2 | I enjoy political participation because I want to have as much of a say in running government as possible. | J'apprécie la participation politique parce que je veux avoir, autant que possible, la possibilité de m'exprimer dans le fonctionnement d'un gouvernement. |
| 3 | I try to avoid situations where someone else tells me what to do. | J'essaie d'éviter les situations dans lesquelles quelqu'un me dit ce que je dois faire. |
| 4 | I would prefer to be a leader than a follower. | Je préfère être un leader plutôt qu'un suiveur. |
| 5 | I enjoy being able to influence the actions of others. | J'apprécie de pouvoir influencer les actions des autres. |
| 6 | I am careful to check everything on an automobile before I leave for a long trip. | Je fais attention à tout vérifier dans une voiture avant de partir pour un long voyage. |
| 7* | Others usually know what is best for me. | Les autres savent généralement ce qui est bon pour moi. |
| 8 | I enjoy making my own decisions. | J'apprécie de pouvoir prendre mes propres décisions. |
| 9 | I enjoy having control over my own destiny. | J'apprécie d'avoir le contrôle sur ma propre destinée. |
| 10* | I would rather someone else took over the leadership role when I'm involved in a group project. | Je préfère que quelqu'un d'autre prenne le rôle de leader quand je suis impliqué(e) dans un projet de groupe. |
| 11 | I consider myself to be generally more capable of handling situations than others are. | Je me considère comme généralement plus capable de gérer les situations que les autres. |
| 12 | I'd rather run my own business and make my own mistakes than listen to someone else's orders. | Je préfère diriger ma propre affaire et faire mes propres erreurs qu'écouter les ordres de quelqu'un d'autre. |
| 13 | I like to get a good idea of what a job is all about before I begin. | J'aime avoir une idée globale et claire d'un travail avant de le commencer. |

| | | |
|---|---|---|
| 14 | When I see a problem I prefer to do something about it rather than sit by and let it continue. | Quand je rencontre un problème, je préfère faire quelque chose à propos de celui-ci plutôt que de rester passif. |
| 15 | When it comes to orders, I would rather give them than receive them. | En ce qui concerne les ordres, je préfère en donner qu'en recevoir. |
| 16* | I wish I could push many of life's daily decisions off on someone else. | J'aimerais pouvoir me décharger du poids des décisions du quotidien sur quelqu'un d'autre. |
| 17 | When driving, I try to avoid putting myself in a situation where I could be hurt by someone else's mistake. | Sur la route, j'essaie d'éviter de me mettre dans des situations où je pourrais être blessé à cause de quelqu'un d'autre. |
| 18 | I prefer to avoid situations where someone else has to tell me what is I should be doing. | Je préfère éviter les situations dans lesquelles quelqu'un doit me dire ce que je devrais faire. |
| 19* | There are many situations in which I would prefer only one choice rather than having to make a decision. | Dans de nombreuses situations, je préfère n'avoir qu'une seule option plutôt que de devoir faire un choix entre plusieurs options. |
| 20* | I like to wait and see if someone else is going to solve a problem so that I don't have to be bothered by it. | J'aime attendre de voir si quelqu'un d'autre va résoudre un problème de sorte que je n'ai pas à m'en soucier. |
| *Reverse item. | | |

## The NARS questionnaire for evaluating the negative attitude toward robots (study 2, section 2.3)

The order of the questions follows the original questionnaire proposed by Nomura *et al.* [NOR 06]. The left column reports the original questions in English. The right column reports our translation of the questions in French.

## Three subtasks designed to test the trust in functional savvy (study 2, section 2.3)

### *Evaluating sounds*

The experimenter explained that the goal of the evaluation was to determine the most acute sounds between two. The experimenter would use the computer on her right to play the two sound stimuli consecutively, saying "first" and "second" before each sound (in French, "le premier" and "le deuxième"). The sound stimuli were sinusoidal waveforms at a pure frequency, lasting 1.5 sec, and were generated by the Audacity software.

| No. | Questionnaire item in English | Questionnaire item in French | Subscale |
|---|---|---|---|
| 1 | I would feel uneasy if robots really had emotions. | Je me sentirais mal à l'aise si les robots avaient réellement des émotions. | S2 |
| 2 | Something bad might happen if robots developed into living beings. | Quelque chose de mauvais pourrait se produire si les robots devenaient des êtres vivants. | S2 |
| 3 | I would feel relaxed talking with robots. | Je serais détendu(e) si je parlais avec des robots. | S3* |
| 4 | I would feel uneasy if I was given a job where I had to use robots. | Je me sentirais mal à l'aise dans un travail où je devrais utiliser des robots. | S1 |
| 5 | If robots had emotions, I would be able to make friends with them. | Si les robots avaient des émotions, je serai capable de devenir ami(e) avec eux. | S3 |
| 6 | I feel comforted being with robots that have emotions. | Je me sens réconforté(e) par le fait d'être avec des robots qui ont des émotions. | S3* |
| 7 | The word "robot" means nothing to me. | Le mot ''robot'' ne signifie rien pour moi. | S1 |
| 8 | I would feel nervous operating a robot in front of other people. | Je me sentirais nerveux/nerveuse de manœuvrer un robot devant d'autres personnes. | S1 |
| 9 | I would hate the idea that robots or artificial intelligences were making judgments about things. | Je détesterais que les robots ou les intelligences artificielles fassent des jugements sur des choses. | S1 |
| 10 | I would feel very nervous just standing in front of a robot. | Le simple fait de me tenir face à un robot me rendrait très nerveux/nerveuse. | S1 |
| 11 | I feel that if I depend on robots too much, something bad might happen. | Je pense que si je dépendais trop fortement des robots, quelque chose de mauvais pourrait arriver. | S2 |
| 12 | I would feel paranoid talking with a robot. | Je me sentirais paranoïaque de parler avec un robot. | S1 |
| 13 | I am concerned that robots would be a bad influence on children. | Je suis préoccupé(e) par le fait que les robots puissent avoir une mauvaise influence sur les enfants. | S2 |
| 14 | I feel that in the future society will be dominated by robots. | Je pense que dans le futur la société sera dominée par les robots. | S2 |
| *Reverse item. | | | |

The frequencies of the sounds were as follows:

1) (“equal” or “50–50” case): 450 and 450 Hz;

2) (“different” case): 117 and 200 Hz;

3) (“ambiguous” case): 450 and 455 Hz;

4) (“ambiguous” case): 100 and 110 Hz.

For each question, the experimenter asked “Which is the most acute sound: the first or the second?”. We remark that for most of the participants, this was the most difficult task as the environmental conditions were probably not optimal for evaluating the sounds (background noise from computers, etc.).

### *Evaluating images*

The experimenter explained that the goal of this evaluation was to determine the dominant color in an image. Three colors were suggested for each image. The experimenter would use the computer on her right to show the images consecutively. The images are shown in the figure.

For each image, the suggested colors were as follows:

1) (“equal” or “50–50” case): abstract pattern; possible colors: blue, yellow, black. Here, the robot always answered blue or yellow, depending on the participant’s answer;

2) (“different” case): a red ladybird on a green leaf; possible colors: green, yellow, red. Here, the robot always chooses “green” as the most dominant color;

3) (“ambiguous” case): industrial scenario; possible colors: green, blue, gray. Here, the robot always answered green or blue, depending on the participant’s answer;

4) (“ambiguous” case): gray worm on a green leaf; possible colors: green, gray, black. Here, the robot always answered green or gray, depending on the participant’s answer.

For each image, the experimenter asked “Which is the dominant color in this image: the <first color>, the <second color> or the <third color>?”.

## Evaluating weights

The experimenter explained that the goal of this evaluation was to determine the heavier object between two. Four pairs of bottles, with different or similar shape, color and weight (filled or not) were used. The bottles are shown in Figure 2.4.

For each pair, the weights are as follows:

1) (“equal” or “50–50” case): two white bottles filled with big beans, identical in appearance and weight (42 g);

2) (“different” case): two bottles filled with big beans, with different appearance and different weight (41 and 73 g);

3) (“ambiguous” case): two bottles filled with small beans, with similar appearance and slightly different weight (50 and 63 g);

4) (“ambiguous” case): two bottles filled with big beans, with different appearance and slightly different weight (65 and 57 g).

For each pair of bottles, the experimenter asked “Which bottle is heaviest: the first or the second?”. The experimenter always took the bottle with number one in her left hand, and number two in her right hand. She gave the bottles to the participant, who could take his/her time to evaluate, sometimes swapping the bottles in the hands. Then, the experimenter gave the bottles to the robot. The operator would activate a predesigned arm trajectory, opening the hands. The experimenter put the bottles in the middle of the palm, saying “Close the hands”; the robot would then grasp the bottles with a predesigned grasp (see Figure ). The experimenter repeated the question. Given the different sizes of the bottles, two different preprogrammed grasps were used. When the robot had said “first” or “second”, the experimenter said “open the hands”, retrieved the bottles and gave them back to the participant for getting his/her final decision.

# Appendix 5

Questionnaire completed by pupils at the end of the RObeeZ project: self-evaluation of cognitive, affective, social and meta-cognitive impacts of the project.

Plusieurs questions te demandent de colorier ton ressenti: *(Legend to assess your feedback about the project:)*

| | |
|---|---|
| ☆ | Pas du tout *(Not at all)* |
| ☆☆ | Un peu *(A little)* |
| ☆☆☆ | Moyennement *(More or less)* |
| ☆☆☆☆ | Beaucoup *(A lot)* |
| ☆☆☆☆☆ | Complètement *(Completely)* |

1) As-tu utilisé des connaissances que tu avais déjà sur les abeilles et sur les robots ?

*(Did you use any pre-existing knowledge about bees and robots?)*

2) As-tu corrigé des connaissances que tu avais déjà sur les abeilles ?

*(Did you correct any of your existing knowledge about bees?)*

3) As-tu corrigé des connaissances que tu avais déjà sur les robots ?

*(Did you correct any of your existing knowledge of robots*?)

4) As-tu appris des choses nouvelles sur les abeilles ? Si oui, lesquelles ?

*(Did you learn anything new about bees? If so, what?)*

5) As-tu appris des choses nouvelles sur les robots ? Si oui, lesquelles ?
*(Did you* learn *anything new about robots? If so, what?)*

6) As-tu cherché des informations sur les abeilles et le robot :

*(Did you have to seek out extra information about bees and robots by either:)*

– Aidé par la maitresse *(Asking the teacher)*

– Seul *(By yourself)*

– Avec les élèves de ton groupe ? (*Asking the other pupils in your group?*)

7) Le sujet principal du projet a été propose
*(The main subject was given by:)*

– Par la maitresse *(The teacher)*

– Par vous, les élèves *(You, the pupils)*

8) Après avoir observé des vraies abeilles et des photos d'abeilles, es-tu passé au dessin:
*(After observing bees and photos of bees did you then draw:)*

– En suivant les instructions de la maitresse *(According to the teacher's instructions)*

– En inventant le dessin toi-même *(By inventing your own drawing)*

– En collaboration avec les élèves de ton groupe ? (*By collaborating with the other pupils in your group?*)

9) Après avoir dessiné les abeilles, es-tu passé à la construction :
*(After drawing the bees did you then begin construction:)*

– En suivant un modèle sur le manuel (*By following the instructions in the manual*)

– En suivant les instructions de la maitresse *(By following the teacher's instructions)*

– En l'inventant toi-même (*By inventing it yourself*)

– En inventant avec les élèves de ton groupe ? (*By inventing it with the pupils in your group?*)

10) Après avoir observé des vraies abeilles et des vidéos d'abeilles, as-tu schématisé leur comportement :
*(After having observed the bees and watched videos of bees, did you begin modeling their behaviour:)*

– En suivant les instructions de la maitresse *(By following the teacher's instructions)*

– En identifiant toi-même différents types de comportement pour différents types d'abeilles *(By identifying bee behavior patterns by yourself)*

– En collaborant avec les élèves de ton groupe ? (*By collaborating with the other pupils in your group?*)

11) Comment as-tu programmé ton robot abeille ?
*(How did you program your bee robot?)*

– En imitant des programmes trouvés sur un manuel ou sur le web *(By imitating programs from the manual or from the web)*

– En essayant des blocs de programmes que l'enseignant te proposait *(By using the programming blocks proposed by the teacher)*

– En inventant ton propre programme *(By inventing your own programming)*

– En collaborant avec les élèves de ton groupe (*By collaborating with the pupils in your group*)

12) As-tu communiqué avec les autres pendant la construction du robot-abeille ?
*(During the construction of the robot bee, did you communicate with:)*

– Avec les élèves de ton groupe *(The pupils in your group)*

– Avec les élèves des autres groupes *(The pupils of other groups?)*

13) As-tu communiqué avec les autres pendant la programmation du robot-abeille ?
*(During the programming of the robot bee, did you communicate with:)*

– Avec les élèves de ton groupe *(The pupils in your group)*

– Avec les élèves des autres groupes *(The pupils of other groups?)*

14) Penses-tu que ton équipe a fait un bon travail ?
*(Do you think that your team did a good job?)*

15) Penses-tu que les équipes de ta classe se sont bien cordonnés ?
*(Do you think that the teams in your class coordinated well together?)*

16) Penses-tu que l'étude des abeilles a été utile pour mieux comprendre le fonctionnement du robot ?
*(Do you think that observing the bees helped you create the robot?)*

17) Penses-tu que la construction et la programmation des robots a été utile pour mieux comprendre la vie des abeilles ?
*(Do you think that constructing the robot helped you understand the life and the behavior of the bees?)*

18) As-tu aimé ce projet ?
*(Did you enjoy this project?)*

19) Si tu voulais faire un autre projet, saurais tu remettre en place les différentes étapes ?
*(Would you be able to repeat these steps if you were to experience this project again?)*

# Bibliography

[ACK 91] ACKERMANN E., "The agency model of transactions: toward an understanding of children's theory of control", in HAREL I., PAPERT S. (eds), *Constructionism*, Norwood, NJ: Ablex Publisher, pp. 367–379, 1991.

[ACK 01] ACKERMANN E., "Constructivisme et constructionnisme: quelle difference?", in *Constructivisme: usages et perspectives en education*, SRED, Genève: Département de l'Instruction Publique, Service de la recherche en education, vol. I & II, pp. 85–97, 2001.

[ADA 90] ADAMS D., *Le Guide du Routard Galactique*, Paris: Denoël, 1990.

[AHL 03] AHLGREN D.J., VERNER I.M., "An international view of robotics as educational medium", *International Conference on Engineering Education 2002*, Manchester, UK, 2003.

[ALA 92] ALAVI M., JOACHIMSTHALER E.A., "Revisiting DSS implementation research: a meta-analysis of the literature and suggestions for researchers", *MIS Quarterly*, vol. 16, no. 1, pp. 95–116, 1992.

[ALB 04] ALBERTA, *Learning and Teaching Resources Branch*, Focus on inquiry: a teacher's guide to implementing inquiry-based learning, http://www.learning.gov.ab.ca/k_12/curriculum/bySubject/focusoninquiry.pdf, 2004.

[ALI 12] ALIMISIS D., "Robotics in education and education in robotics: Shifting focus from technology to pedagogy", in OBDRZÁLEK D. (ed.), *Proceedings of the 3rd International Conference on Robotics in Education*, Prague: Charles University in Prague, Faculty of Mathematics and Physics, pp. 7–14, 2012.

[ALI 13] ALIMISIS D., "Educational robotics: open questions and new challenges", *Themes in Science & Technology Education*, vol. 6, no. 1, pp. 63–71, 2013.

[ALI 09] ALIMISIS D. (ed.), *TERECoP Project: Teacher Education on Robotics-Enhanced Constructivist Pedagogical Methods*, Athens: School of Pedagogical and Technological Education, ASPETE, 2009.

[ALP 35] ALPORT G.W., "Attitudes", in MURCHISON C. (ed.), *Handbook of Social Psychology*, Worcester, MA: Clark University Press, pp. 798–844, 1935.

[AND 01] ANDERSON L.W., KRATHWOHL D.R., AIRASIAN P.W. *et al.*, *A Taxonomy for Learning, Teaching, and Assessing: a Revision of Bloom's Taxonomy of Educational Objectives*, New York: Longman, 2001.

[AND 96] ANDERSON J.R., REDER L.M., SIMON H.A., "Situative versus cognitive perspectives: form versus substance", *Educational Researcher*, vol. 26, no. 1, pp. 18–21, 1996.

[AND 08] ANDLER D., GUERRY B. (eds), *Apprendre Demain, Sciences Cognitives et Éducation à l'ère Numérique*, Paris: Hatier, pp. 26–51, 2008.

[ARG 94] ARGLE D., " La programmation parallèle à la portée d'enfants de sept ans", in DENIS B., BARON G.L. (eds), *Regards sur la Robotique pédagogique. Actes du quatrième colloque sur la robotique pédagogique*, Liège, Paris: INRP, pp. 221–231,1994.

[ARL 09] ARLEGUI J., FAVA N., MONTFALCON S. *et al.*, "Robotics as learning object", in ALIMISIS, D. (ed.), *TERECoP project, Teacher Education on Robotics Enhanced Constructivist Pedagogical Methods*, Athens: School of Pedagogical and Technological Education, ASPETE, pp. 27–45, 2009.

[ARR 03] ARROYO I.A., ARROYO A.A., SCHWAARTZ E.M., "The integration of robotics in grade school to promote interest in math and science", in TOSUNOGLU S. (ed.), *Proceedings of Florida Conference on Recent Advances in Robotics* (FCRAR), Florida Atlantic University, Boca Raton, Florida, available at http://www.mil.ufl.edu/publications/, 2003.

[ATM 08] ATMATZIDOU S., MARKELIS I., DEMETRIADIS S., "The use of LEGO Mindstorms in elementary and secondary education: game as a way of triggering learning", in CARPIN S., NODA I., PAGELLO E. *et al.* (eds), *Proceedings of International Conference on Simulation Modeling and Programming for Autonomous Robots* (SIMPAR), Venice, Italy, pp. 22–30, 2008.

[AVR 13] AVRIL M., CHETOUANI M., SABOURET N., "Étude d'une collaboration socio-affective entre une personne et le robot Jazz", *Interfaces Numériques*, vol. 2, no. 1, pp. 57–76, 2013.

[BAD 13] BADDOURA R., VENTURE G., "Social vs. useful HRI: experiencing the familiar, perceiving the robot as a sociable partner and responding to its actions", *International Journal of Social Robotics*, vol. 5, pp. 529–547, 2013.

[BAC 77] BACH R., *Illusions. Adventures of the Reluctant Messiah*, Dell Publishing, New York, 1977.

[BAR 96] BARON G.L., BRUILLARD E., *L'informatique et ses usagers dans l'éducation*, Paris: Presses Universitaires de France, 1996.

[BAR 05] BARTNECK C., NOMURA T., KANDA T. *et al.*, "A cross-cultural study on attitudes towards robots", *Paper presented at the International Conference on Human Computer Interaction* (HCI 2007). Las Vegas, USA, 2005.

[BAR 07b] BARKER B.S., ANSORGE J., "Robotics as means to increase achievement scores in an informal learning environment", *Journal of Research on Technology Education*, vol. 39, no. 3, pp. 229–243, 2007.

[BAR 07b] BARTNECK C., SUZUKI T., KANDA T. *et al.*, "The influence of people's cultures and prior experiences with Aibo on their attitude towards robots", *AI & Society*, vol. 21, pp. 217–230, 2007.

[BAR 07c] BARTNECK C., VERBUNT M., MUBIN O. *et al.* "To kill a mockingbird robot", in *Proceedings of 2nd ACM/IEEE International Conference onHuman-Robot Interaction* (HRI) Washington DC, USA, pp. 81–87, 2007.

[BAR 09a] BARAK B., ZADOK Y., "Robotics projects and learning concepts in science, technology and problem solving", *International Journal of Technology and Design Education*, vol. 19, no. 3, pp. 289–307, 2009.

[BAR 09b] BARTLEY J.E., MAYHEW L.M., FINKELSTEIN N.D., "Promoting children's understanding and interest in science through informal education", *Physics Education Research Conference*, vol. 1179, no. 1, pp. 93–96, 2009.

[BAR 09c] BARTNECK C., CROFT E., KULÍC D., "Measurement instruments for the anthropomorphism, animacy, likeability, perceived intelligence, and perceived safety of robots", *International Journal of Social Robotics*, vol. 1, no. 1, pp. 71–81, 2009.

[BAR 09d] BARTOLINI BUSSI M.G., MARIOTTI M.A., "Mediazione semiotica nella didattica della matematica: artefatti e segni nella tradizione di Vygotskij", *L'Insegnamento della Matematica e delle Scienze Integrate*, vol. 32, A-B, pp. 270–294, 2009.

[BEA 01] BEACH R., MYER J., *Inquiry-based English Instruction: engaging students in life and literature*, New York: Teachers College Press, 2001.

[BEE 11] BEER J. M., PRAKASH A., MITZNER T.L. *et al.*, "Understanding robot acceptance", Technical Report HFA-TR-1103, Georgia Institute of Technology School of Psychology – Human Factors and Aging Laboratory, available at: http://hdl.handle.net/1853/39672, 2011.

[BEL 10] BELL S., "Project-Based learning for the 21st century: skills for the future", *The Clearing House*, vol. 83, pp. 39–43, 2010.

[BEN 12] BENITTI F.V.B., "Exploring the educational potential of robotics in schools: A systematic review", *Computers & Education*, vol. 58, no. 3, pp. 978–988, 2012.

[BER 00] BEREITER C., SCARDAMALIA M., "Process and product in problem based learning", in EVENSON D., HMELO C. (eds), *Problem Based Learning: A Research Perspective on Learning Interactions*, Mahwah: Lawrence Erlbaum Associates, pp. 185–195, 2000.

[BER 11] BERAN T.N., RAMIREZ-SERRANO A., KUZYK R. *et al.*, "Would children help a robot in need?", *International Journal of Social Robotics*, vol. 3, no. 1, pp. 83–92, 2011.

[BER 02] BERS M., PONTE I., JUELICH K. *et al.*, "Teachers as designers: integrating robotics in early childhood education", *Information Technology in Childhood Education Annual*, 1, Association for the Advancement of Computing in Education (AACE), pp. 123–145, 2002.

[BER 08a] BERNSTEIN D., CROWLEY K., "Searching for signs of intelligent life: An investigation of young children's beliefs about robot intelligence", *Journal of the Learning Sciences*, vol. 17, no. 2, pp. 225–247, 2008.

[BER 08b] BERS M.U., *Blocks to robots: learning with technology in the early childhood classroom*, New York: Teachers College Press, 2008.

[BLA 99] BLASCOVICH J., MENDES W.B., HUNTER S.B. *et al.*, "Social "facilitation", as challenge and threat", *Journal of Personality and Social Psychology*, vol. 77, pp. 68–77, 1999.

[BIR 04] BIROS D., DALY M., GUNSCH G., "The influence of task load and automation trust on deception detection", *Group Decision and Negotiation*, vol. 13, pp. 173–189, 2004.

[BOL 10] BONNEL B., *Vive la robolution*, Paris: Editions JC Lattès, 2010.

[BOT 02] BOTTINO R., CHIAPPINI G., "Advanced technology and learning environments: their relationship within the Arithmetic problem-solving domain", in ENGLISH L., BARTOLINI BUSSI M., JONES G. *et al.* (eds), *Handbook of International Research in Mathematics Education*, Mahwah, NJ: Erlbaum, pp. 757–786, 2002.

[BOU 14] BOUCENNA S., NARZISI A., TILMONT E. *et al.*, *Cognitive Computation*, vol. 6, no. 4, pp. 722–740, 2014.

[BRA 10] BRANGIER E., HAMMES-ADELÉ S., BASTIEN J.-M.C., "Analyse critique des approches de l'acceptation des technologies de l'utilisabilité à la symbiose humain-technologie-organisation", *Revue européenne de psychologie appliqué*, vol. 60, pp. 129–146, 2010.

[BRE 03] BREAZEAL C., "Emotion and sociable humanoid robots", *International Journal of Human-Computer Studies*, vol. 59, pp. 119–155, 2003.

[BRE 10] BREDENFELD A., LEIMBACH T., "The Roberta Initiative", *Proceedings of Workshops International Conference on Simulation, Modeling and Programming for Autonomous Robots* (SIMPAR), Darmstadt, Germany, pp. 558–567, 2010.

[BRI 07] BRIDGEMAN N.C., BRIDGEMAN E.K., "Education: a role for robots?", Paper presented at the *20th Annual Conference of the National Advisory Committee on Computing Qualifications*, Nelson, New Zealand, 2007.

[BRO 78] BROWN J.S., BURTON R.R., "Diagnostic models for procedural bugs in basic mathematical skills", *Cognitive Science*, vol. 2, pp. 155–192, 1978.

[BRU 02] BRU M., DONNAY J., *Recherches, pratiques et savoirs en education*, Bruxelles: De Boeck, 2002.

[BRU 61] BRUNER J.S., "The act of discovery", *Harvard Educational Review*, vol. 31, no. 1, pp. 21–32, 1961.

[BUR 79] BURGER J.M., COOPER H.M., "The desirability of control", *Motivation and Emotion*, vol. 3, no. 4, pp. 381–393, 1979.

[BUR 07] BURKE C.S., SIMS D.E., LAZZARA E.H. *et al.*, "Trust in Leadership: A multi-level review and integration", *The Leadership Quarterly*, vol. 18, pp. 606–632, 2007.

[CAC 02] CACI B., CARDACI M., LUND H.H., Assessing Educational Robotics by the "Robot Edutainment Questionnaire", Technical Report, The Maersk McKinley Moeller Institute for Production Technologies, University of Southern Denmark, 2002.

[CAC 05] CACI, B., D'AMICO A., "Robotics: a new tool for education of subjects with cognitive diseases", in CHIAZZESE G., ALLEGRA M., CHIFARI A. *et al.* (eds), *Methods and Technologies for Learning*, WIT Transaction on Information and Communication Technologies, Southampton, UK: WIT Press, vol. 34, pp. 563–567, 2005.

[CAR 04] CARBONARO M., REX M., CHAMBERS J., "Using Lego robots to promote problem solving in a project based learning environment", *Interactive Multimedia Electronic Journal of Computer Enhanced Learning*, vol. 6, no. 1, 2004.

[CAT 12] CATLIN D., BLAMIRES M., "The principles of educational robotic applications (ERA): a framework for understanding and developing educational robots and their activities", *Advancing Education* (Summer), pp. 2050–1331, 2012.

[CAT 10] CATLIN D., BLAMIRES M., "The principles of educational robotic applications (ERA): A framework for understanding and developing educational robots and their activities", Paper Presented at *Constructionism*, Paris, France, available at: http://www.naace.co.uk/1948, 2010.

[CHA 07] CHAMBERS J., CARBONARO M., REX M., "Scaffolding knowledge construction through robotic technology: A middle school case study", *Electronic Journal for the Integration of Technology in Education*, vol. 6, pp. 5–70, 2007.

[CHA 10] CHANG C.W., LEE J.H., CHAO P.Y. *et al.*, "Exploring the possibility of using humanoid robots as instructional tools for teaching a second language in primary school", *Educational Technology & Society*, vol. 13, no. 2, pp. 13–24, 2010.

[CHA 02] CHARLIER B., PÉRAYA D. (eds), *Technologie et Innovation en pédagogie: Dispositifs Innovants de Formation pour l'enseignement Supérieur*, Bruxelles: De Boeck, 2002.

[CHE 13] CHENG C.C., HUANG P.L., HUANG K.H., "Cooperative learning in Lego robotics projects: exploring the impacts of group formation on interaction and achievement", *Journal of Networks*, vol. 8, no. 7, pp. 1529–1535, 2013.

[CHO 15] CHOI H., KIM K., "A Study on the students' perception change for the characteristics of the teacher before and after robot-based instruction in the middle school", *Advanced Science and Technology Letters*, vol. 92, pp. 58–61, 2015.

[CHR 10] CHRONAKI A., ALIMISI R., "Engaging young children to "control" technology: emotion, negotiation, agency", in MENEGATTI E. (ed.), *Proceedings of International Conference on Simulation, Modeling and Programming for Autonomous Robots* (SIMPAR), Darmstadt, Germany, pp. 624–638, 2010.

[CHU 91] CHUNG J., "Collaborative learning strategies: The design of instructional environments for the emerging new school", *Educational Technology*, vol. 31, no. 12, pp. 15–22, 1991.

[CLE 87] CLEMENTS D.H., "Componential employment and development in LOGO programming environments", in *Biennial Meetings of the Society for Research in Child Development*, Baltimore, Maryland, 1987.

[CLE 99] CLEMENTS, D., "The future of educational computing research: the case of computer programming", *Information Technology in Child Education*, vol. 1, pp. 147–179, 1999.

[CLE 93] CLEMENTS D.H., MEREDITH J.S., "Research on logo: effects and efficacy", *Journal of Computing in Childhood Education*, vol. 4, pp. 263–290, 1993.

[COE 12] COECKELBERGH M., "Can we trust robots?", *Ethics and Information Technology*, vol. 14, no. 1, pp. 53–60, 2012.

[CON 94] CONWAY M., PAUSCH R., GOSSWEILER, R. *et al.*, "Alice: a rapid prototyping system for building virtual environments", in ADELSON B., DUMAIS S., OLSON J.S. (eds), *Proceedings of the Human Factors in Computing Systems Conference*, Boston, Massachusetts: ACM Press, pp. 295–296, 1994.

[CRI 12] CRISP R.J., TURNER R.N., "Imagined intergroup contact", in HODSON G. HEWSTONE M. (eds), *Advances in Intergroup Contact*, Hove: Psychology Press, Taylor & Francis, pp. 135–151, 2012.

[CSI 10] CSINK K., FARKAS L., "Life-long playing instead of life-long learning: teaching robotics without robots and computers", in MENEGATTI E. (ed.), *Proceedings of International Conference on Simulation, Modeling and Programming for Autonomous Robots* (SIMPAR), Radovljica, Slovenia, Univerza v Ljubljani, Pedagoška fakulteta, pp. 439–448, 2010.

[CUE 05] CUEVAS P., LEE O., HART J. *et al.*, "Improving science inquiry with elementary students of diverse backgrounds", *Journal of Research in Science Teaching*, vol. 42, no. 3, pp. 337–357, 2005.

[DAU 07] DAUTENHAHN K., "Socially intelligent robots: dimensions of human–robot interaction", *Philosophical Transaction of the Royal Society B*, vol. 362, no. 1480, pp. 679–704, 2007.

[DAV 03] DAVIDSON J.E., STERNBERG R.J., *The Psychology of Problem Solving*, Cambridge University Press, 2003.

[DAV 89] DAVIS F., BAGOZZI R., WARSHAW R., "User acceptance of computer technology: a comparison of two theoretical models", *Management Science*, vol. 35, pp. 982–1003, 1989.

[DEB 13] DE BRUIJN M.L.E., The Base of Trust in Human-Robot Interaction, Thesis, Radboud University Nijmegen, 2013.

[DEC 96] DE CORTE E., "Changing views of computer supported learning environments for the acquisition of knowledge and thinking skills", in VOSNIADOU S., DE CORTE E., GLASER R. *et al.* (eds), *International Perspectives on the Design of Technology-Supported Learning Environments*, Mahwah: Lawrence Erlbaum Associates, pp. 129–145, 1996.

[DEF 07] DE FREITAS S., JARVIS S., "Serious games – engaging training solutions: A research and development project for supporting training needs", *British Journal of Educational Technology*, vol. 38, no. 3, pp. 523–525, 2007.

[DEG 13] DE GRAAF M.M.A., ALLOUCH S.B., "Exploring influencing variables for the acceptance of social robots", *Robotics and Autonomous Systems*, vol. 61, pp. 1476–1486, 2013.

[DEJ 05] DE JONG F., KOLLÖFFEL B., VAN DER MEIJDEN H. *et al.*, "Regulative processes in individual, 3D and computer-supported cooperative learning contexts", *Computers in Human Behavior*, vol. 21, pp. 645–670, 2005.

[DEM 08] DE MICHELE M.S., DEMO G.B., SIEGA S., "A Piedmont schoolnet for a k-12 mini-robots programming project: experiences in primary schools", in *Proceedings of the TERECoP Workshop "Teaching with robotics"*, SIMPAR Conference, Venice, Italy, pp. 90–99, 2008.

[DEM 12] DEMO G., MORO M., PINA A. *et al.*, "In and out of the school activities implementing IBSE and constructionist learning methodologies by robotics", in BRADLEY BARKER S., GWEN N., GRANDGENETT N. *et al.* (eds), *Robots in K-12 Education*, IGI Global, pp. 66–92, 2012.

[DEN 94] DENIS B., BARON G.L. (eds), "Regards sur la Robotique pédagogique", *Actes du Quatrième Colloque sur la Robotique Pédagogique*, Liège, Paris: INRP, 1994.

[DEN 01] DENIS B., HUBERT S., "Collaborative learning in an educational robotics environment", *Computers in Human Behavior*, vol. 17, pp. 465–480, 2001.

[DER 05] DE RUYTER B., SAINI P., MARKOPOULOS P. *et al.*, "Assessing the effects of building social intelligence in a robotic interface for the home", *Interacting with Computers*: *Social Impact of Emerging Technologies*, vol. 17, no. 5, pp. 522–541, 2005.

[DES 09] DESAI M., STUBBS K., STEINFELD A. *et al.*, "Creating trustworthy robots: lessons and inspirations from automated systems", *Proceedings of the AISB Convention: New Frontiers in Human-Robot Interaction*, Edinburgh, Scotland, 2009.

[DES 10] DESCOLA P., *La Fabrique des Images, Visions du Monde et Formes de la Représentation*, Paris: Somogy/Musée du quai Branly, 2010.

[DES 15] DESTEPHE M., BRANDAO M., KISHI T. *et al.*, "Walking in the uncanny valley: importance of the attractiveness on the acceptance of a robot as a working partner", *Frontiers in Psychology*, vol. 6, no. 204, 2015.

[DES 12] DESTENO D., BREAZEAL C., FRANK R.H. *et al.*, "Detecting the trustworthiness of novel partners in economic exchange", *Psychological Science*, vol. 23, pp. 1549–1556, 2012.

[DET 06] DETTORI G., GIANNETTI T., PAIVA A. *et al.* (eds), *Technology Mediated Narrative Environment for Learning*, Rotterdam: Sense Publishers, 2006.

[DIG 08] DIGNATH C., BÜTTNER G., LANGFELDT H.P., "How can primary school students learn self-regulated learning strategies most effectively?, A meta-analysis on self-regulation training programmes", *Educational Research Review*, vol. 3, pp. 101–129, 2008.

[DEV 01] DE VRIES E., "Les logiciels d'apprentissage: panoplie ou éventail", *Revue française de pédagogie*, vol. 137, pp. 105–116, 2001.

[DIL 99] DILLENBOURG P., "What do you mean by collaborative leraning?", in DILLENBOURG P. (ed.), *Collaborative-learning: Cognitive and Computational Approaches*, Oxford: Elsevier, pp.1–19, 1999.

[DIL 01] DILLON A., "User acceptance of information technology", in KARWOWSKI W. (ed.), *Encyclopedia of Human Factors and Ergonomics*, London: Taylor and Francis, 2001.

[DIN 14] DINET J., VIVIAN R., "Exploratory investigation of attitudes towards assistive robots for future users", *Le travail humain*, vol. 77, pp. 105–125, 2014.

[DIN 15a] DINET J., "Would you be friends or colleagues with a robot?", Oral presentation, at CIRHUS (CNRS UMI 3199, New York University). available at: http://www.researchgate.net/publication/274069148_Would_you_be_friends_or_colleagues_with_a_robot, 2015.

[DIN 15b] DINET J., VIVIAN R., "Perception et attitudes à l'égard des robots anthropomorphes en France: validation d'une échelle d'attitudes", *Psychologie française* vol. 60, pp. 173–189, 2015.

[DRI 94] DRIVER R., ASOKO H., LEACH, J. *et al.*, "Constructing scientific knowledge in the classroom", *Educational Researcher*, vol. 23, no. 7, pp. 5–12, 1994.

[DRU 00] DRUIN A., HENDLER J. (eds), *Robots for Kids: Exploring New Technologies for Learning*, San Diego: Academic Press, 2000.

[DUF 03] DUFFY B., "Anthropomorphism and the social robot", *Robotics and Autonomous Systems*, vol. 42, pp. 177–190, 2003.

[DUF 10] EGUCHI A., "What is educational robotics?, Theories behind it and practical implementation", in GIBSON D., DODGE B. (eds), *Proceedings of Society for Information Technology & Teacher Education International Conference*, Chesapeake: AACE, pp. 4006–4014, 2010.

[EGU 12] EGUCHI A., URIBE L., "Educational robotics meets inquiry-based learning: integrating inquiry-based learning into educational robotics", in LENNEX L., NETTLETON K. (eds), *Cases on Inquiry through Instructional Technology in Math and Science*, New York: IGI Publishing, pp. 327–366, 2012.

[ENG 08] ENGEL S.L., RANDALL K., "How teachers respond to children's inquiry", *American Education Research Journal*, vol. 46, no. 1, pp. 183–202, 2008.

[ESH 07] ESHACH H., "Bridging in-school and out-of-school learning: formal, non-formal, and informal education", *Journal of Science Education and Technology*, vol. 16, PP. 171–190, 2007.

[FIS 11] FISCHER K., "Interpersonal variation in understanding robots as social actors", in *Proceedings of International Conference on Human-Robot Interaction*, Lausanne, Switzerland, pp. 53–60, 2011.

[FLA 87] FLAVELL J.H., FLAVELL E.R., GREEN F.L., "Young children's knowledge about the apparent-real and pretend-real distinctions", *Developmental Psychology*, vol. 23, no. 6, pp. 816–822, 1987.

[FLA 79] FLAVELL J.H., "Metacognition and cognitive monitoring: A new area of cognitive-developmental inquiry", *American Psychologist*, vol. 34, no. 10, pp. 906–911, 1979.

[FON 03] FONG T., NOURBAKHSH I., DAUTENHAHN K., "A survey of socially interactive robots", *Robotics and Autonomous Systems*, vol. 42, pp. 143–166, 2003.

[FOR 06] FORD M., DACK G., PREJEAN L., "Robotics: Implementing problem based learning in teacher education and field experience", in CRAWFORD C., CARLSEN R., MCFERRIN K. *et al.* (eds), *Proceedings of Society for Information Technology and Teacher Education International Conference*, Chesapeake: AACE, pp. 3410–3416, 2006.

[FRA 08] FRANGOU S., PAPANIKOLAOU K., ARAVECCHIA L. *et al.*, "Representative examples of implementing educational robotics in school based on the constructivist approach", in CARPIN S., NODA I., PAGELLO E. *et al.* (eds), *Proceedings of International Conference on Simulation Modeling and Programming for Autonomous Robots* (SIMPAR), Venice, Italy, pp. 54–65, 2008.

[FRI 13] FRIESEN S., SCOTT D., Inquiry-based learning: a review of the research literature, Calgary, AB: Galileo Education Network, available at: https://inspiring.education.alberta.ca/wp-content/uploads/2014/04/Inquiry-Based-Learning-A-Review-of-the-Research-Literature.pdf, 2013.

[FRI 14] FRIDIN M., BELOKOPYTOV M., "Acceptance of socially assistive humanoid robot by preschool and elementary school teachers", *Computers in Human Behavior*, vol. 33, pp. 23–31, 2014.

[FUM 12] FUMAGALLI M., IVALDI S., RANDAZZO M. *et al.*, "Force feedback exploiting tactile and proximal force/torque sensing, theory and implementation on the humanoid robot iCub", *Autonomous Robots*, vol. 33, no. 4, pp. 381–398, 2012.

[GAG 09] GAGNÉ P.P., LEBLANC N., ROUSSEAU A., *Apprendre... Une Question de Stratégies: Développer les Habiletés Liées aux Fonctions Exécutives*, Montréal: Chenelière, 2009.

[GAL 92] GALLAGHER S., STEPIEN W., ROSENTHAL H., "The effects of problem-based learning on problem solving", *Gifted Child Quarterly*, vol. 36, no. 4, pp. 195–200, 1992.

[GAL 04] GALILEO EDUCATIONAL NETWORK, *What is inquiry?*, available at http://www.galileo.org/inquiry-what.html, 2004.

[GAT 03] GATT S., VELLA Y. (eds), *Constructivist Teaching in Primary School: Examples in Social Studies, Science, Mathematics, Design and Technology and ICT*, Malta: Agenda Publishers, 2003.

[GAT 16] GATT S., MESTDAGH N., VAN DE KEERE K. *et al.*, The Pri-Sci-Net project: bringing inquiry at primary school, 2016.

[GAU 11a] GAUDIELLO I., TIJUS C., ZIBETTI E., "Du schéma d'action à la symbolisation: programmer des tâches robotiques pour favoriser l'apprentissage de l'arithmétique élémentaire", *Actes du 53ème Congrès National de la Société Française de Psychologie: La pratique: un lieu de théories*, Metz, France, pp. 128–129, 2011.

[GAU 11b] GAUDIELLO I., ZIBETTI E., TIJUS C., "Educational robotics in science education: why and how, A frame to exploit the salient features of robotics kits to teach and learn science", *Educational Technology Research and Development, Workshop How Science Works and how to Teach it*, Aarhus, Denmark, 2011.

[GAU 12a] GAUDIELLO I., ZIBETTI E., PINAUD C.A., "Control heuristics for educational robots: a pilot study", *Proceedings of 3rd International Workshop Teaching Robotics, Teaching with Robotics, Integrating Robotics in School Curriculum*, Riva del Garda, Italy, pp. 67–75, 2012.

[GAU 12b] GAUDIELLO I., ZIBETTI E., "Using control heuristics as a means to explore the educational potential of robotics kits", *Themes in Science and Technology Education*, vol. 6, no. 1, pp. 15–28, 2012.

[GAU 14] GAUDIELLO I., ZIBETTI E., "La robotique éducationnelle: état des lieux et perspectives, *Psychologie Française*, vol. 58, pp. 17–40, 2014.

[GAU 15] GAUDIELLO I., LEFORT S., ZIBETTI E., "The place of robots in common-sense ontology and their prospected educative roles: how firm our representations are?", *Computers in Human Behavior*, vol. 50, pp. 259–273, 2015.

[GAU 16a] GAUDIELLO I., ZIBETTI E., "Are we ready to robots? A critical survey on the notion of acceptance", *in press*, 2016.

[GAU 16b] GAUDIELLO I., ZIBETTI E., "Educational robotics as a semiotic tool for learning", *in press*, 2016.

[GEL 03] GELMAN S.A., *The Essential Child: Origins of Essentialism in Everyday Thought*, New York, Oxford University Press, 2003.

[GEL 04] GELMAN S.A.,"Psychological essentialism in children", *Trends in Cognitive Science*, vol. 8, no. 9, pp. 404–409, 2004.

[GIL 04] GILLIES M., BALLIN D., "Integrating autonomous behavior and user control for believable agents", *Third International Joint Conference on Autonomous Agents and Multiagent Systems*, New York, USA, pp. 336–343, 2004.

[GOL 01] GOLAN R., KYZA E.A., REISER B.J. *et al.*, "Scaffolding the task of analyzing animal behavior with the animal landlord software", *Annual Meeting of the American Educational Research Association*, New Orleans, LA, 2001.

[GOL 04] GOLDMAN R., EGUCHI A., SKLAR E., "Using educational robotics to engage inner-city students with technology", in KAFAI Y., SANDOVAL W., ENYEDY N. *et al.* (eds), *Proceedings of Sixth International Conference of the Learning Sciences* (ICLS), Mahwah, NJ: Lawrence Erlbaum, pp. 214–221, 2004.

[GRI 12] GRIMAUD, E., VIDAL D., "Aux frontières de l'humain, pour une anthropologie comparée des créatures artificielles", *Gradhiva*, vol. 15, pp. 4–25, 2012.

[GRO 07] GROOM V., NASS C., "Can robots be teammates? Benchmarks in human-robot teams", *Interaction Studies*, vol. 8, no. 3, pp. 483–2500, 2007.

[GUI 99] GUIMELLI, C., *La pensée sociale*, Paris: Presses Universitaires de France, 1999.

[HAC 99] HACKER D.J., Metacognition: "Definitions and empirical foundations", *The MIT Encyclopedia of Cognitive Sciences*, MIT Press, 1999.

[HAE 04] HAEFNER L.A., ZEMBAL-SAUL C., "Learning by doing? prospective elementary teachers' developing understandings of scientific inquiry and science teaching and learning", *International Journal of Science Education*, vol. 26, no. 13, pp. 1653–1674, 2004.

[HAM 06a] HAMILL L., "Controlling smart devices in the home", *The Information Society*, vol. 22, no. 4, pp. 241–249, 2006.

[HAM 06b] HAMILL L., HARPER R., "Talking intelligence: a historical and conceptual exploration of speech-based human-machine interaction in smart homes", in *Proceedings of the International Symposium on Intelligent Environments* (ISIE), Cambridge, UK, Cambridge: MSR Press, pp. 121–128, 2006.

[HAN 09] HAN J., KIM D., "r-Learning services for elementary school students with a teaching assistant robot", *Proceedings of 4th ACM/IEEE International Conference on Human-Robot Interaction*, La Jolla, California, USA, pp. 255–256, 2009.

[HAN 11] HANCOCK P.A., BILLINGS D.R., SCHAEFER K.E. *et al.*, "A meta-analysis of factors affecting trust in human-robot interaction", *Human Factors*, vol. 53, no. 5, pp. 517–27, 2011.

[HAR 00] HARACKIEWICZ J. (eds), *Intrinsic Motivation: Controversies and New Directions*, San Diego: Academic Press, 2000.

[HAR 04] HARLEN W., *Evaluating Inquiry-Based Science Developments*, "Paper commisioned by the national research council in preparation for a meeting on the status of evaluation of inquiry-based science education", "University of cambridge and university of bristol, available at: http://www7.nationalacademies.org/bose/WHarlen_inquiry_Mtg_paper.pdf, 2004.

[HEE 10a] HEERINK M., Assessing acceptance of assistive social robots by aging adults, PhD thesis, University of Amsterdam, 2010.

[HEE 10b] HEERINK M., KRÖSE B.J.A., WIELINGA B.J. *et al.*, "Assessing acceptance of assistive social agent technology by older adults: The almere model", *International Journal of Social Robotics*, vol. 2, pp. 361–375, 2010.

[HER 09] HEERINK M., KROSE B., WIELINGA B. *et al.*, "Influence of social presence on acceptance of an assistive social robot and screen agent by elderly users", *Advanced Robotics*, vol. 23, pp. 1909–1923, 2009.

[HEU 94] HEUDIN J.C., *La Vie Artificielle*, Paris: Hermès Science, 1994.

[HIM 12] HIMBERT M.N., *Le Robot Pensant*, Paris: Editions du Moment, 2012.

[HIN 04] HINDS P. J., ROBERTS T. L., JONES H., "Whose job is it anyway? a study of human–robot interaction in a collaborative task", *Human-Computer Interaction*, vol. 19, pp. 151–181, 2004.

[HME 07] HMELO-SILVER C.E., DUNCAN R.G., CLARK A., "Scaffolding and achievement in problem-based and inquiry learning: a response to kirschner, sweller, and clark (2006)", *Educational Psychologist*, vol. 42, no. 2, pp. 99–107, 2007.

[HME 04] HMELO-SILVER C.E., "Problem-based learning: what and how do students learn?", *Educational Psychology Review*, vol. 16, pp. 235–266, 2004.

[HO 10] HO C.C., MACDORMAN K., "Revisiting the uncanny valley theory: developing and validating an alternative to the godspeed indices", *Computers in Human Behavior*, vol. 26, no. 6, pp. 1508–1518, 2010.

[HOF 05] HOFMANN W., GAWRONSKI B., GSCHWENDNER T. *et al.*, "A meta-analysis on the correlation between the implicit association test and explicit self-report measures", *Personality and Social Psychology Bulletin*, vol. 31, pp. 1369–1385, 2005.

[HOF 09] HOFFMAN R.R., LEE J.D., WOODS D.D. *et al.*, "The dynamics of trust in cyberdomains", in HOFFMAN R.R., BRADSHAW J.M., FORD K.M. (eds), *Human-Centered Computing*, IEEE Computer Society, pp. 5–11, 2009.

[HOR 11] HORN M. S., CROUSER R.J., BERS M.U., "Tangible interaction and learning: The case for a hybrid approach", *Personal and Ubiquitous Computing, Special Issue on tangibles and children*, vol. 16, no. 4, pp. 379–389, 2011.

[HOS 08] HOSKINS B., FREDRIKSSON U., "Learning to learn: What is it and can it be measured", JRC Scientific and Technical Report EUR 23432 EN, Ispra: European Commission, 2008.

[HRO 14] HROMADA D., GAUDIELLO I., "Robot introduction to moral induction model and its deployment in artificial agents", in SEIBT J., HAKLI R., NORSKOV M. (eds), *Social Robots and the Future of Social Relations*, Amsterdam: IOS Press, pp. 209–2016, 2014.

[HSU 07] HSU S.H., CHOU C.Y., CHEN F.C. *et al.*, "An investigation of the differences between robot and virtual learning companions' influences on students' engagement", in CHAN T.W., PAIVA A., WILLIAMSON SHAFFER D. *et al.* (eds), *Proceedings of the First IEEE International Workshop on Digital Game and Intelligent Toy Enhanced Learning*, Jhongli, Taiwan, pp. 41–48, 2007.

[HUG 04] HUGHES P., MORE A.J., WILLIAMS M., *Aboriginal Ways of Learning*, Adelaide: Paul Hughes, 2004.

[HUS 06] HUSSAIN S., LINDH J., SHUKUR G., "The effect of LEGO training on pupils' school performance in mathematics, problem solving ability and attitude: Swedish Data", *Educational Technology & Society*, vol. 9, no. 3, pp. 182–194, 2006.

[HYU 08] HYUN E. J., KIM S.Y., JANG S.K. *et al.*, "Comparative study of effects of language education program using intelligence robot and multimedia on linguistic ability of young children", in *Proceedings of the 17th IEEE International Workshop on Robot and Human Interactive Communication* (RO-MAN), Munich, Germany, pp. 187–192, 2008.

[ILI 10] ILIEVA V., "Robotics in primary school, how to do it?", in MENEGATTI E. (ed.), *Proceedings of International Conference on Simulation, Modeling and Programming for Autonomous Robots* (SIMPAR), Darmstadt, Germany, pp. 596–605, 2010.

[ION 07] IONITA S., IONITA A.I., "Steps towards a constructionist methodological approach" *Proceedings of Eco Media International Conference*pp, Pitesti, Romania, pp. 119–125, 2007.

[IVA 13] IVALDI S., ANZALONE S.M., ROUSSEAU W. *et al.*, "Cues for making a humanoid child more "human-like' during social learning tasks", *Proceedings of International Workshop Towards social humanoid robots: What makes interaction human-like?* (IROS), Tokyo, Japan, pp. 1–6, 2013.

[IVA 14a] IVALDI S., NGUYEN S.M., LYUBOVA N. *et al.*, "Object learning through active exploration", *IEEE Transactions on Autonomous Mental Development*, vol. 6, pp. 56–72, 2014.

[IVA 14b] IVALDI S., ANZALONE S.M., ROUSSEAU W. *et al.*, "Robot initiative in a team learning task increases the rhythm of interaction but not the perceived engagement", *Frontiers in Neurorobotics*, vol. 8, no. 5, pp. 1–23, 2014.

[JAC 13] JACK A.I., DAWSON A.J., BEGANY K.L. *et al.*, "fMRI reveals reciprocal inhibition between social and physical cognitive domains", *NeuroImage*, vol. 66, pp. 385–401, 2013.

[JEW 11] JEWELL S.J., The Effects of the NXT robotics curriculum on high school students' attitudes in science based on grade, gender, and ethnicity, PhD thesis, Liberty University, 2011.

[JIP 07] JIPSON J., GELMAN S., "Robots and rodents: Children's inferences about living and nonliving kinds", *Child Development*, vol. 78, 1675–1688, 2007.

[JOH 05] JOHNSON G.M., "Instructionism and constructivism: Reconciling two very good ideas", *International Journal of Special Education*, vol. 24, no. 3, pp. 90–98, 2005.

[JOO 13] JOOSSE M., SARDAR A., LOHSE M. *et al.*, "BEHAVE-II, A revised set of measures to assess users' attitudinal and behavioral responses to a robot's social hehaviors", *International Journal of Social Robotics*, vol. 5, no. 3, pp. 379–388, 2013.

[JON 06] JONASSEN D.H., *Modeling with Technology, Mindtools for Conceptual Change* (3rd ed.), Columbus: Pearson Prentice Hall, 2006.

[JUN 09] JUNG U., LEE E., LEE Y., "Educational robot research trends analysis", available at http://citeseerx.ist.psu.edu/viewdoc/download?doi=10.1.1.372.9609&rep=rep1&type=pdf, 2009.

[KAH 06] KAHN P.H. JR., FRIEDMAN B., PEREZ-GRANADOS D.R. *et al.*, "Robotic pets in the lives of preschool children", *Interaction Studies: Social Behavior and Communication in Biological and Artificial Systems*, vol. 7, pp. 405–436, 2006.

[KAH 11] KAHN H.P. JR., REICHERT A.L., GARY H.E. *et al.*, "The new ontological category hypothesis", *Proceedings of Human–Robot Interaction (HRI)*, Lausanne, Switzerland, pp. 159–160, 2011.

[KAH 12] KAHN JR., PETER H., KANDA T. *et al.*, "Robovie, you'll have to go into the closet now: Children's social and moral relationships with a humanoid robot", *Developmental Psychology*, vol. 48, no. 2, pp. 303–314, 2012.

[KAH 09] KAHN P.H. JR., SEVERSON R.L., RUCKERT J.H., "The human relation with nature and technological nature", *Current Directions in Psychological Sciences*, vol. 18, no. 1, pp. 37–42, 2009.

[KAH 03] KAHNEMAN D., "A perspective on judgment and choice: Mapping, bounded rationality", *American Psychologist*, vol. 58, pp. 697–720, 2003.

[KAL 95] KALISH C.W., "Essentialism and graded membership in animal and artifact categories", *Memory and Cognition*, vol. 23, no. 3, pp. 335–353, 1995.

[KAM 14] KAMIDE H., MAE Y., TAKUBO T. *et al.*, "Direct comparison of psychological evaluation between virtual and real humanoids; personal space and subjective impressions, *International Journal of Human-Computer Studies*, vol. 72, no. 5, pp. 451–459, 2014.

[KAN 03] KANG J.P., "A study on the education of robot in elementary school", *Journal of Society of Korean Practical Arts Education*, vol. 16 no. 4, pp. 97–113, 2003.

[KAN 04a] KANDA T., ISHIGURO H., IMAI M. *et al.*, "Development and evaluation of interactive humanoid robots", *Proceedings of the IEEE*, vol. 92, no. 11, pp. 1839–1850, 2004.

[KAN 04b] KANDA T., HIRANO T., EATON D. *et al.*, "Participation of interactive humanoid robots in human society", *Journal of the Robotics Society of Japan*, vol. 22, no. 5, pp. 94–105, 2004.

[KAN 07] KANDA T., SATO R., SAIWAKI N. *et al.*, "A two-month field trial in an elementary school for long-term human–robot interaction", *IEEE Transactions on Robotics*, vol. 23, no. 5, pp. 962–971, 2007.

[KAN 12] KANIARASU P., STEINFELD A., DESAI M. *et al.*, "Potential measures for detecting trust changes", *Proceedings of the 7th Annual ACM/IEEE International Conference on Human-Robot Interaction* (HRI), Boston, Massachusetts, pp. 241–242, 2012.

[KAP 05] KAPLAN F., *Les Machines Apprivoisées*, Paris: Vuibert, 2005.

[KAP 04] KAPLAN F., "Who is afraid of the humanoid?, Investigating cultural differences in the acceptance of robots", *International Journal of Humanoid Robotics*, vol. 1, no. 3, pp. 1–16, 2004.

[KEL 55] KELLY G., *The Psychology of Personal Constructs*, New York: Norton, 1955.

[KER 06] KEREPESIA A., KUBINYIB E., JONSSONC G.K. *et al.*, "Behavioural comparison of human–animal (dog) and human–robot (AIBO) interaction", *Behavioural Processes*, vol. 73, no. 1, pp, 92–99, 2006.

[KET 06] KETELHUT D.J., DEDE C.C.J., Assessing Inquiry Learning, Paper presented at the National Association of Research in Science Teaching, San Francisco, CA, 2006.

[KID 03] KIDD C.D., *Sociable Robots:* The role of presence and task in human-robot interaction, Master's thesis, Massachusetts Institute of Technology, Boston, Massachusetts, 2003.

[KII 07] KIILI K., "Foundation for problem-based gaming", *British Journal of Educational Technology*, vol. 38, no. 3, pp. 394–404, 2007.

[KIM 14] KIM K., BAEK J., "A Study on the Teachers' perception of school curriculum", Implementation about Robot-based Education in Korea, *Advanced Science and Technology Letters*, vol. 59, pp. 105–108, 2014.

[KIR 06] KIRSCHNER P.A., SWELLER J., CLARK R.E., "Why Minimal guidance during instruction does not work: An analysis of the failure of constructivist, discovery, problem-based, experiential, and inquiry-based teaching", *Educational Psychologist*, vol. 41, no. 2, pp. 75–86, 2006.

[KLA 88] KLAHR D., CARVER S.M., "Cognitive objectives in a LOGO debugging curriculum: instruction, learning, and transfer", *Cognitive Psychology*, vol. 20, pp. 362–404, 1988.

[KLA 00] KLAHR D., *Exploring Science: The Cognition and Development of Discovery Processes*. Cambridge: MIT Press, 2000.

[KOA 14] KOAY K.L., SYRDAL D.S., ASHGARI-Oskoei M. *et al.*, "Social roles and baseline proxemic preferences for a domestic service robot", *International Journal of Social Robotics*, vol. 6, no. 4, pp. 469–488, 2014.

[KOL 03] KOLODNER J.L., CAMP P.J., CRISMOND D. *et al.*, "Problem-based learning meets case-based reasoning in the middle-school science classroom: putting learning by design™ into practice", *The Journal of The Learning Sciences*, vol. 12, no. 4, pp. 495–547, 2003.

[KOM 11] KOMATSU T., YAMADA S., "Adaptation gap hypothesis: how differences between users' expected and perceived agent functions affect their subjective impression", *Journal Systemics, Cybernetics and Informatics*, vol. 9, no. 1, pp. 67–74, 2011.

[KNO 97] KNORR-CETINA K., "Sociality with objects: social relations in postsocial knowledge societies", *Theory Culture Society*, vol. 14, no. 4, pp. 1–30, 1997.

[KRO 00] KRÖSE B., BOGAARD R., HIETBRINK N., "Programming robots is fun: RoboCup jr", in VAN DEN BOSCH A., WEIGAND H. (eds), *Proceedings of Belgium-Netherlands AI Conference* (BNAIC), De Efteling, Kaatsheuvel, pp. 29–36, 2000.

[KUC 12] KUCHENBRANDT D., EYSSEL F., "The mental simulation of a humanrobot interaction: positive effects on attitudes and anxiety toward robots", *Proceedings of the 21st IEEE International Symposium in Robot and Human Interactive Communication* (RO-MAN), Paris, France, pp. 463–468, 2012.

[KUL 07] KULÍC D., CROFT E., "Affective state estimation for human-robot interaction", *IEEE Transactions on Robotics*, vol. 23, no. 5, pp. 991–1000, 2007.

[KUM 04] KUMAR A.N., "Three years of using robots in an artificial intelligence course: lessons learned", *Journal on Educational Resources in Computing*, vol. 4, no. 3, pp. 1–16, 2004.

[KYN 08] KYNIGOS C., "Black-and-white-box perspectives to distributed control and constructionism in learning with robotics", in CARPIN S., NODA I., PAGELLO E. *et al.* (eds), *Proceedings of International Conference on Simulation Modeling and Programming for Autonomous Robots* (SIMPAR), Venice, Italy, pp. 41–49, 2008.

[LAF 74] LAFFERTY J.C., EADY P.M., *The Desert Survival Problem*, Plymouth: Experimental Learning Methods, 1974.

[LAL 13] LALLÉE S., HAMANN K., STEINWENDER J. *et al.*, "Cooperative human robot interaction systems: IV, communication of shared plans with naïve humans using gaze and speech", *Proceedings of the IEEE/RSJ International Conference on Intelligent Robots and Systems* (IROS), Tokyo, Japan, pp. 129–136, 2013.

[LAP 10] LA PAGLIA F., CACI B., LA BARBERA D. *et al.*, "Using robotics construction kits as metacognitive tools: a research in an italian primary school", *Studies in Health Technology and Information*, vol. 154, pp. 110–4, 2010.

[LAU 99] LAU K., TAN H., ERWIN B. *et al.*, "Creative learning in school with LEGO programmable robotics products", *Proceedings of AAEE/IEEE Conference on Frontiers in Education*, Champaign, Stipes Publishing, L.L.C, 1999.

[LEC 87] LECLERCQ D., L'ordinateur et le défi de l'apprentissage. Horizon, Philips Professional Systems, vol. 13, pp. 29–32, 1987.

[LEC 95] LECLERCQ P., Approche technologique de l'education et de la formation, thesis, University of Liège, 1995.

[LEE 04] LEE J.D., SEE K.A., "Trust in automation: designing for appropriate reliance", *Human Factors: The Journal of the Human Factors and Ergonomics Society*, vol. 46, no. 1, pp. 50–80, 2004.

[LEE 07] LEE E., LEE Y., "The effect of a robot programming learning on problem solving ability", *The Journal of Korean Association of Computer Education*, vol. 10, no. 6, pp. 1–9, 2007.

[LEE 11] LEE N., SHIN H., SUNDAR S.S., "Utilitarian vs hedonic robots, role of parasocial tendency and anthropomorphism in shaping user attitudes", *Proceedings of the 6th International Conference on Human-Robot Interaction*, Lausanne, Switzerland, pp. 183–184, 2011.

[LEE 13a] LEE J.J., KNOX W.B., WORMWOOD J.B. *et al.*, "Computationally modeling interpersonal trust", *Frontiers in Psychology*, vol. 4, pp. 1–14, 2013.

[LEE 13b] LEE K., SULLIVAN A., BERS M.U., "Collaboration by design: Using robotics to foster social interaction in kindergarten", *Computers in the Schools*, vol. 30, no. 3, pp. 271–281, 2013.

[LEM 13] LE MAITRE J., CHETOUANI M., "Self-talk Discrimination in human-robot interaction situations for supporting social awareness", *International Journal of Social Robotics*, vol. 5, no. 2, pp. 277–289, 2013.

[LEN 05] LE NY J.F., *Comment L'esprit Produit Du Sens: Notions et Résultats des Sciences Cognitive*, Paris: Odile Jacob, 2005.

[LEP 00] LEPPER M.R., HENDERLONG J., "Turning "play" into "work" and "work" into "play": 25 years of research on intrinsic versus extrinsic motivation", in SANSONE C., HARACKIEWICZ J.M. (eds), *Intrinsic and Extrinsic Motivation: The Search for Optimal Motivation and Performance*, San Diego, CA: Academic Press, pp. 25–307, 2000.

[LEV 15] LEVILLAIN F., ZIBETTI E., "Moving on its own: how do audience interacts with an autonomous moving artwork", *Proceedings of Computer-Human Interaction* (CHI), Seoul, Korea, pp. 695–702, 2015.

[LEV 16] LEVILLAIN F., ZIBETTI E., LEFORT S., "Interpreting the behavior and interacting with non-anthropomorphic robotic artworks", *in press*, 2016.

[LEV 08] LEVY S.T., MIODUSER D., "Does it "want" or "was it programmed to?" Kindergarten children's explanations of an autonomous robot's adaptive functioning", *International Journal of Technology and Design Education*, vol. 18, no.4, pp. 337–359, 2008.

[LI 10] LI D., RAU P.L.P., LI Y., "A cross-cultural-study: effects of robot appearance and task", *International Journal of Social Robotics*, vol. 2, no. 2, pp. 175–186, 2010.

[LIM 94] LIMBOS B., "L'utilisation de représentations graphiques en robotique pédagogique", in DENIS D., BARON G.L. (eds), *Regards sur la Robotique Pédagogique, ACTES du Quatrième Colloque sur la Robotique Pédagogique*, Liège. Paris: INRP, pp. 65–83, 1994.

[LIN 07] LINDH J., HOLGERSSON T., "Does lego training stimulate pupils' ability to solve logical problems?", *Computers and Education*, vol. 49, no. 4, pp. 1097–1111, 2007.

[LIN 85] LINN M.C., "The cognitive consequences of programming", *Instruction in Classrooms Educational Researcher*, vol. 14, no. 5, pp. 14–29, 1985.

[LOC 04] LOCKERD A., BREAZEAL C., "Tutelage and socially guided robot learning", *Proceedings of Intelligent Robots and System*, vol. 4, pp. 3475–3480, 2004.

[LOH 10] LOHSE M., Investigating the influence of situations and expectations on user behavior – empirical analyses in human-robot interaction, PhD thesis, Bielefeld University, 2010.

[LOO 10] LOOIJE R., NEERINCX M.A., CNOSSEN F., "Persuasive robotic assistant for health self-management of older adults: design and evaluation of social behaviors", *International Journal of Human-Computer Studies*, vol. 68, no. 6, pp. 386–397, 2010.

[LU 11] LU C.M., KANG S., HUANG S.C. *et al.*, "Building student understanding and interest in science through embodied experiences with LEGO robotics", in BASTIAENS T., EBNER M. (eds), *Proceedings of EdMedia: World Conference on Educational Media and Technology*, Association for the Advancement of Computing in Education (AACE), pp. 2225–2232, 2011.

[LUB 10] LUBART T.I., PACTEAU C., JAQUET A.Y. *et al.*, "Children's creative potential: An empirical study of measurement issues", *Learning and Individual Differences*, vol. 20, pp.388–392, 2010.

[LUN 00] LUND H.H., PAGLIARINI L., "RoboCup Jr. with Lego Mindstorms", *Proceedings of International Conference on Robotics and Automation* (ICRA), San Francisco CA, USA, pp. 813–819, 2000.

[LUN 03a] LUND H.H., NIELSEN J., "An edutainment robotics survey", in MURASE K., ASAKURA T. (eds), *Dynamic Systems Approach for Embodiment and Sociality*, *International Series on Advanced Intelligence*, vol. 6, pp. 173–205, 2003.

[LUN 03b] LUND H.H., "Adaptive robotics in the entertainment industry", *Proceedings of IEEE International Symposium on Computational Intelligence in Robotics and Automation* (ISCIRA), Kobe, Japan, pp. 595–602, 2003.

[LUN 07] LUNSFORD E., MELEAR C.T., ROTH W.M. *et al.*, "Proliferation of inscriptions and transformations among pre-service science teachers engaged in authentic science", *Journal of Research in Science Teaching*, vol. 44, no. 4, pp. 538–564, 2007.

[MAC 09] MACDORMAN K.F., VASUDEVAN S.K., HO C.C., "Does Japan really have robot mania? Comparing attitudes by implicit and explicit measures", *AI and Society*, vol. 23, no, 4, pp. 485–510, 2009.

[MAL 15] MALLE B.F., SCHEUTZ M., VOIKLIS J., "Sacrifice one for the good of many?, people apply different moral norms to human and robot agents", in *Proceedings of the 10th Annual ACM/IEEE International Conference on Human-Robot Interaction* ACM, Portland, USA, pp. 117–124, 2015.

[MAR 00] MARTIN F., MIKHAK B., RESNICK M. *et al.*, "To Mindstorms and Beyond: Evolution of a Construction Kit for Magical Machines", in DRUIN A., HENDLER J. (eds), *Robots for Kids: Exploring New Technologies for Learning Experiences*, San Francisco: Morgan Kaufman Academic Press, 2000.

[MAR 04a] MARBLE J., L.DAVID J., BRUEMMER D. *et al.*, "Evaluation of Supervisory vs. peer-peer interaction with human-robot teams", paper presented at the *37th Annual Hawaii International Conference on System Sciences*, Hawaii, 2004.

[MAR 04b] MARX R.M., BLUMENFELD P.C., KRAJCIK J.S. *et al.*, "Inquiry-Based Science in the Middle Grades: Assessment of Learning in Urban Systemic Reform", *Journal of Research in Science Teaching*, vol. 41, no. 10, pp. 1063–1080, 2004.

[MAT 04] MATARIC M.J., "Robotics Education for All Ages", Paper presented at *AAAI Spring Symposium on Accessible, Hands-on AI and Robotics Education*, Palo Alto, CA, 2004.

[MAY 95] MAYER R.C., DAVIS J.H., SCHOORMAN F.D., "An integrative model of organizational trust", *Academy of Management Review*, vol. 20, pp. 709 –734, 1995.

[MAY 02a] MAYER R.E., "A taxonomy for computer-based assessment of problem solving", *Computers in Human Behavior*, vol. 18, pp. 623–632, 2002.

[MAY 02b] MAYER R.E., *The Promise of Educational Psychology: Teaching for Meaningful Learning*, Upper Saddle River: Prentice Hall, 2002.

[MCB 10] McBRIDE M., MORGAN S., "Trust calibration for automated decision aids", Institute for Homeland Security Solutions, Research brief, available at: http://sites.duke.edu/ihss/files/2011/12/McBride_Research_Brief.pdf, pp. 1–11, 2010.

[MCD 12] McDONALD S., HOWELL J., "Watching, creating and achieving: Creative technologies as a conduit for learning in the early years", *British Journal of Educational Technology*, vol. 43, no. 4, pp. 641–651, 2012.

[MCL 06] McLOUGHLIN C.E., LUCA J., Applying situated learning theory to the creation of learning environments to enhance socialization and self-regulation, Edith Cowan University: Ecu Publications, available at: http://ro.ecu.edu.au/cgi/viewcontent.cgi?article=3270&context=ecuworks, 2006.

[MCL 10] McLOUGHLIN C., LEE M.J.W., "Personalised and self-regulated learning in the Web 2.0 era: International exemplars of innovative pedagogy using social software", *Australasian Journal of Educational Technology*, vol. 26, no. 1, pp. 28–43, 2010.

[MEL 09] MELTZOFF A.N., KUHL P.K., MOVELLAN J. *et al.*, "Foundations for a new science of learning", *Science*, 19, vol. 325, no. 5938, pp. 284–288, 2009.

[MER 98] MERRIAM S.B., *Qualitative Research and Case Study Applications in Education*, San Francisco: Jossey-Bass, 1998.

[MER 08] MERRITT S.M., ILGEN D.R., "Not all trust is created equal: dispositional and history-based trust in humanautomation interactions", *Human Factors*, vol. 50, no. 2, pp. 194–210, 2008.

[MER 12] MERRITT S.M., HEIMBAUGH H., LACHAPELL J. *et al.*, "I trust it, but I don't know why: Effects of implicit attitudes toward automation on trust in an automated system", *Human Factors: The Journal of the Human Factors and Ergonomics Society*, vol. 55, no. 3, pp. 520–534, 2012.

[MES 08] MESTDAGH N., GATT S., KOSACK W. *et al.*, *The Pillars of Effective Learning in Science*, Berlin: Mensch and buch verlag, 2008.

[MEY 05] MEYER J.A., GUILLOT A., GIRARD B. *et al.*, "The Psikharpax project: towards building an artificial rat", *Robotics and Autonomous Systems*, vol. 50, no. 4, pp. 211–223, 2005.

[MIK 13] MIKROPOULOS T.A., BELLOU J., "Educational robotics as mindtools", *Themes in Science & Technology Education*, vol. 6, no. 1, pp. 5–14, 2013.

[MIN 10] MINNER D.D., LEVY A.J., CENTURY J., "Inquiry-based science instruction – what is it and does it matter?, Results from a research synthesis years 1984–2002", *Journal of Research in Science Teaching*, vol. 47, no. 4, pp. 474–496, 2010.

[MIO 96] MIODUSER D., VENEZKY R.L., GONG B., "Students' perceptions and designs of simple control systems", *Computers in Human Behavior*, vol. 12, no. 3, pp. 363–388, 1996.

[MIO 09] MIODUSER D., LEVY S.T., Tallis V., "Episodes to scripts to rules: Concrete-abstractions in kindergarten children's explanations of a robot's behavior", *International Journal of Technology and Design Education*, vol. 19, no. 1, pp. 15–36, 2009.

[MIT 08] MITNIK R., RECABARREN M., NUSSBAUM M. *et al.*, "Collaborative robotic instruction: A graph teaching experience", *Computers & Education*, vol. 53, no. 2, pp. 330–342, 2008.

[MON 14] MONROE A.E., DILLON K.D., MALLE B.F., "Bringing free will down to Earth: People's psychological concept of free will and its role in moral judgment", *Consciousness and Cognition*, vol. 27, pp. 100–108, 2014.

[MON 08] MONTAGUE R.P., "Free will", *Current Biology*, vol. 18, pp. R584–R585, 2008.

[MOR 00] MORITA T., IWATA H., SUGANO S., "Human symbiotic robot design based on division and unification of functional requirements", *Proceedings of IEEE International Conference on Robotics and Automation* (ICRA), San Francisco, CA, USA, vol. 3, pp. 2229–2234, 2000.

[MOR 06] MORK S.M., "viten.no – digital teaching programs in science education", *NorDiNa*, vol. 3, pp. 84–8, 2006.

[MOS 05] MOSHKINA L., ARKIN R.C., "Human perspective on affective robotic behavior: a longitudinal study", *Proceedings of IEEE International Conference on Intelligent Robots and Systems* (IROS), Edmonton, Canada, pp. 1444–1451, 2005.

[MOV 09] MOVELLAN J., ECKHARDT M., VIRNES M. *et al.*, "Sociable robot improves toddler vocabulary skills", *Proceedings of the 4th ACM/IEEE International Conference on Human Robot Interaction*, La Jolla, California, USA, pp. 307–308, 2009.

[MUB 13] MUBIN O., STEVENS C.J., SHAHID S. *et al.*, "A review of the applicability of robots in education", *Technology for Education and Learning*, vol. 19, pp. 148–163, 2013.

[MUM 11] MUMM J., MUTLU B., "Human-robot proxemics: physical and psychological distancing in human-robot interaction", *Proceedings of 6th ACM/IEEE International Conference on Human Robot Interaction*, Lausanne, Switzerland, pp. 33–338, 2011.

[MUN 01] MUNAKATA Y., "Graded representations in behavioral dissociations", *Trends in Cognitive Sciences*, vol. 5, pp. 309–315, 2001.

[MUR 10] MURPHY R.R., "Human-robot interaction", *Robotics & Automation Magazine*, vol. 17, no. 2, pp. 85–89, 2010.

[NAS 93] NASS C.I., STEUER J., TAUBER E.R. *et al.*, "Anthropomorphism, agency and ethopoeia: computers as social actors", *Computer-Human Interaction Conference*, Amsterdam, The Netherlands, 1993.

[NAS 94] NASS C.I., STEUER J.S., HENRIKSEN L. *et al.*, "Machines and social attributions: performance assessment of computers subsequent to "self-" of "other-" evaluations", *International Journal of Human–Computer Studies*, vol. 40, pp. 543–559, 1994.

[NAS 95] NASS C., MOON Y., FOGG B.J. *et al.*, "Can computer personalities be human personalities?", *International Journal of Human-Computer Studies*, vol. 43, no, 2, pp. 223–239, 1995.

[NAS 96] NASS C., FOGG B.J., MOON Y., "Can computers be teammates?", *International Journal of Human-Computer Studies*, vol. 45, no. 6, pp. 669–678, 1996.

[NAS 97] NASS C.I., MOON Y., MORKES J. *et al.*, "Computers are social actors: a review of current research", in FRIEDMAN B. (ed.), *Human Values and the Design of Computer Technology*, Cambridge: Cambridge University Press, pp. 136–162, 1997.

[NAS 00] NASS C., MOON Y., "Machines and mindlessness: Social responses to computers", *Journal of Social Issues*, vol. 56, no. 1, pp. 81–103, 2000.

[NAS 05] NASS C., BRAVE S., *Wired for Speech: How Voice Activates and Advances the Human-Computer Relationship*. Cambridge: MIT Press, 2005.

[NIC 11] NICHOLS S., "Experimental philosophy and the problem of free will", *Science*, vol. 331, pp. 1401–1403, 2011.

[NOM 14] NOMURA T., "Influences of experiences of robots into negative attitudes toward robots", *Proceedings of the 23rd IEEE International Symposium on Robot and Human Interactive Communication* (RO-MAN), Edinburgh, Scotland, UK, pp. 460–464, 2014.

[NOM 06] NOMURA T., KANDA T., SUZUKI T., "Experimental investigation into influence of negative attitudes toward robots on human-robot interaction", *AI & Society*, vol. 20, pp. 138–150, 2006.

[NON 04] NONAKA S., INOUE K., ARAI T. *et al.*, "Evaluation of human sense of security for coexisting robots using virtual reality, 1st report: evaluation of pick and place motion of humanoid robots", *Proceedings of the IEEE International Conference on Robotics and Automation* (ICRA), New Orleans, vol. 3, pp. 2770–2775, 2004.

[NOU 05] NOURBAKHSH I., HAMNER E., DUNLAVEY B. *et al.*, "Educational results of the personal exploration rover museum exhibit", *Proceedings of the IEEE International Conference on Robotics and Automation*, Barcelona, Spain, pp. 4278–4283, 2005.

[NUG 08] NUGENT G., BARKER B., GRANDGENETT N., "The effect of 4-H robotics and geospatial technologies on science, technology, engineering, and mathematics learning and attitudes", in LUCA J., WEIPPL E. (eds), *Proceedings of World Conference on Educational Multimedia, Hypermedia and Telecommunications*, Chesapeake: AACE, pp. 447–452, 2008.

[NUG 10] NUGENT G., BARKER B., GRANDGENETT N. *et al.*, "Impact of robotics and geospatial technology interventions on youth STEM learning and attitudes", *Journal of Research on Technology in Education*, vol. 42, no. 4, pp. 391–408, 2010.

[OEC 08] OECD, New millennium learners: a project in progress optimising learning: implications of learning sciences research, available at: www.oecd.org/dataoecd/39/51/40554230.pdf, 2008.

[OEC 10] OECD, What students know and can do, PISA 2009 Results, OECD Publishing, 2010.

[OES 06] OESTREICHER L., EKLUNDH K.S., "User expectations on human-robot co-operation", *Proceedings of the 15th IEEE International Symposium on Robot and Human Interactive Communication* (RO-MAN), Hatfield, UK, pp. 91–96, 2006.

[OKI 06] OKITA S.Y., SCHWARTZ D.L., "Young children's understanding of animacy and entertainment robots", *International Journal of Humanoid Robotics*, vol. 3, pp. 393–412, 2006.

[OSB 08] OSBORNE J.F., DILLON J., *Science Education in Europe: Critical Reflections*, London: Nuffield Foundation, 2008.

[OST 69] OSTROM T.M., "The relationship between the affective, behavioral, and cognitive components of attitudes", *Journal of Experimental Social Psychology*, vol. 5, pp. 12–30, 1969.

[PAP 72] PAPERT S., SOLOMON C., "Twenty things to do with a computer", *Educational Technology*, vol. 12, no. 4, pp. 9–18, 1972.

[PAP 80] PAPERT S., *Mindstorms: Children, Computers, and Powerful Ideas*. New York: Basic Books, 1980.

[PAP 81] PAPERT S., "Computer-Based Microworlds as Incubators for Powerful Ideas", in TAYLOR R. (ed.), *The Computer in the School: Tutor, Tool, Tutee*, New York: Teachers College Press, pp. 203–210, 1981.

[PAP 87] PAPERT S., "A critique of technocentrism in thinking about the school of the future", available at: http://www.papert.org/articles/ACritiqueofTechnocentrism.html, 1987.

[PAP 91] PAPERT S., HAREL I., *Situating Constructivism*, New York: Ablex Publishing Corporation, 1991.

[PAP 00] PAPERT S., "The future of school", available at: http://www.papert.org/articles/freire/freirePart2.html, 2000.

[PAR 97] PARASURAMAN R., RILEY V., "Humans and automation: Use, misuse, disuse, abuse", *Human Factors*, vol. 39, pp. 230–253, 1997.

[PAR 01a] PARK, G.R.A. "Review of current status and directon of education of robot and educational materials in elementary schools", *Journal of Korean Practical Arts Education*, vol. 23, no. 3, pp. 323–343, 2001.

[PAR 01b] PARISI D., "Le nuove tecnologie e la scuola: distruzione e ricostruzione", *Sistemi Intelligenti*, vol. 13, pp. 285–298, 2001.

[PAS 13] PASQUINELLI E., "Slippery slopes. Some considerations for favoring a good marriage between education and the science of the mind–brain–behavior, and forestalling the risks", *Trends in Neuroscience and Education*, vol. 2, no. 34, pp. 111–121, 2013.

[PAS 10] PÁSZTOR A., PAP-SZIGETI R., LAKATOS TÖRÖK E., "Effects of using model robots in the education of programming", *Informatics in Education*, vol. 9 no.1, pp. 133–140, 2010.

[PEA 83] PEA R.D., KURLAND D., "Logo programming and the development of planning skill", Techical Report 16, Center for Children & Technology, New York: Bank Street College, 1983.

[PEA 87] PEA R.D., "Cognitive technologies for mathematics education", in SCHOENFELD A. (ed.), *Cognitive Science and Mathematics Education*, pp. 89–122, 1987.

[PER 92] PERKINS D.N., "Transfer of learning", in *International Encyclopedia of Education*, Second Edition Oxford, England: Pergamon Press, available at: http://designprocessineducation.com/design-thinking/tr-ps.htm, 1992.

[PET 04] PETRE M., PRICE B., "Using robotics to motivate back door learning", *Education and Information Technologies*, vol. 9, no. 2, pp. 147–158, 2004.

[PIA 52] PIAGET J., *The Origins of Intelligence in Children*, New York: International University Press, 1952.

[PIA 67] PIAGET J., *Logique et Connaissance Scientifique*, Paris, Gallimard, 1967.

[PIA 72] PIAGET J., *Psychology and Epistemology: Towards a Theory of Knowledge*, Harmondsworth: Penguin, 1972.

[PIA 74] PIAGET J., *To Understand is to Invent*, New York: Basic Books, 1974.

[PIC 05] PICARD R.W., DAILY, S.B., "Evaluating affective interactions: Alternatives to asking what users feel", *CHI Workshop on Evaluating Affective Interfaces*, Portland, Oregon, USA, 2005.

[POP 91] POPPER K.R., *La Connaissance Objective*, Paris: Flammarion, 1991.

[POL 07] POLYA G., *Comment Poser et Résoudre un Problème*, Paris: Jacques Gabay, 2007.

[PRE 01] PRENSKY M., "Digital natives, digital immigrants", *On the Horizon*, vol. 9, no. 5, pp. 1–6, 2001.

[PRE 05] PRENSKY M., "Engage me or enrage me: What today's learners demand", *Educause Review*, vol. 40, no. 5, pp. 61–65, 2005.

[QUI 04] QUINTANA C., REISER B.J., DAVIS E.A. *et al.*, "A scaffolding design framework for software to support science inquiry", *Journal of the Learning Sciences*, vol. 13, pp. 337–386, 2004.

[QUI 99] QUINTANA C., ENG J., CARRA A. *et al.*, "Symphony: A case study in extending learner-centered design through process space analysis", in WILLIAMS M.G., ALTOM M.W., EHRLICH K. (eds), *Proceedings of CHI 99 Conference on Human Factors in Computing Systems*, Reading, MA, Addison-Wesley, pp. 473–480, 1999.

[RAK 11] RAKISONA D.H., YERMOLAYEVAA Y., "How to identify a domain-general learning mechanism when you see one", *Journal of Cognition and Development*, vol. 12, no. 2, pp. 134–153, 2011.

[RAN 02] RANI P., SIMS J., BRACKIN R. *et al.*, "Online stress detection using psychophysiological signals for implicit human-robot cooperation", *Robotica*, vol. 20, pp. 673–685, 2002.

[REG 77] REGAN D.T., FAZIO R., "On the consistency between attitudes and behavior: Look to the method of attitude formation", *Journal of Experimental Social Psychology*, vol. 13, no. 1, pp. 28–45, 1977.

[RES 91] RESNICK M., MARTIN F., "Children and artificial life", in HAREL I., PAPERT S. (eds), *Constructionism*, vol. 10, pp. 379–390, 1991.

[RES 93] RESNICK M., WILENSKY U., "Beyond the deterministic, centralized mindsets: A new thinking for new science", *Annual Meeting of the American Educational Research Association*, Atlanta, GA, 1993.

[RES 96] RESNICK, M., MARTIN F., SARGENT R. *et al.*, "Programmable bricks: toys to think with", *IBM Systems Journal*, vol. 35, nos. 3–4, pp. 443–452, 1996.

[RES 07] RESNICK M., "Sowing the seeds for a more creative society", *International Society for Technology in Education*, pp. 18–22, 2007.

[RES 09] RESNICK M., MALONEY J., MONROY-HERNANDEZ A. *et al.*, "Scratch: programming for all", *Communications of the ACM*, vol. 52, no. 11, pp. 60–67, 2009.

[RIB 96] RIBEIRO C., *RobôCarochinha:* Um Estudo Qualitativo sobre a Robótica Educativa no 1° ciclo do Ensino Básico, Master's thesis, Instituto de Educação e Psicologia da Universidade do Minho, Braga, 1996.

[RIB 11] RIBEIRO C., COUTINHO C., COSTA M.F., "Proposal for the evaluation of educational robotics in basic schools", *Proceedings of International Study Association on Teachers and Teaching (ISATT)*, pp. 831–839, 2011.

[RIB 12] RIBEIRO F., LOPES G., PEREIRA N. *et al.*, "Bot'n roll robotic kit as a learning tool for youngsters", available at http://hdl.handle.net/1822/25747, 2012.

[RIB 09] RIBEIRO C., MACHADO C., COSTA M.F.M. *et al.*, "Robotics as a tool to increase the motivation levels in problematic students", *6th International Conference on Hands-on Science Science for All*, Quest for Excellence, Antalya, Turkey, 2009.

[RIE 08] RIEK L.D., ROBINSON P., "Robot, rabbit, or red herring? Societal acceptance as a function of classification ease", *17th International IEEE Symposium on Robot and Human Interactive Communication, Workshop Robots as Social Actors: Evaluating Social Acceptance and Societal Impact of Robotic Agents, (RO-MAN)*, Munich, Germany, 2008.

[RIE 12] RIEK L.D., "Wizard of Oz studies in HRI: a systematic review and new reporting guidelines", *Journal of Human-Robot Interaction*, vol. 1, no. 1, pp. 119–136, 2012.

[RIE 16] RIETDIJK W., BYRNE J., GAUDIELLO G. *et al.*, "How does the IBSE approach enhance student engagement in science learning?, The case of five European countries", *in Press* 2016.

[RIE 13] RIETHER N., On the Profoundness and Preconditions of Social Responses Towards Social Robots, Experimental Investigations Using Indirect Measurement Techniques, PhD thesis, Universität Bielefeld, Germany, 2013.

[RIN 02] RINGSTAFF C., KELLEY L., *The Learning Return on Our Educational Technology Investment*, San Francisco: WestEd, 2002.

[ROB 05] ROBINSON M., "Robotics-driven activities: can they improve middle school science learning?", *Bulletin of Science, Technology & Society*, vol. 25, no. 1, pp. 73–84, 2005.

[ROC 07] ROCARD M., Science education NOW: a renewed pedagogy for the future of Europe, Brussels: report, European Commission, available at http://ec.europa.eu/research/science-society/document_library/pdf06/report-rocard-onscience-education_en.pdf, 2007.

[ROG 95] ROGERS E.M., *Diffusion of Innovations* (4th ed.), New York: Free Press, 1995.

[ROG 04] ROGERS C., PORTSMORE M., "Bringing engineering to elementary school", *Journal of STEM Education*, vol. 5, nos. 3–4, pp. 17–28, 2004.

[ROS 01] ROSENBERG M.J., *E-learning: Strategies for Delivering Knowledge in the Digital Age*, New York: McGraw-Hill, 2001.

[ROT 71] ROTTER J.B., "Generalized expectancies for interpersonal trust", *American Psychologist*, vol. 26, pp. 443–452, 1971.

[RUG 16] RUGGIERI F., "Les enfants de 5-6 ans apprennent à Nao la notion d'inclusion de classe", Mémoire de Master II en Psychologie Cognitive, Université Paris VIII, 2016.

[RUS 08] RUSK N., RESNICK M., BERG R. *et al.*, "New pathways into robotics: strategies for broadening participation", *Journal of Science Education and Technology*, vol. 17, no. 1, pp. 59–69, 2008.

[RYL 12] RYLANDER J., "Using levels of inquiry in the classroom", in LENNEX L.C., FLETCHER NETTLETON K. (eds), *Cases on Inquiry Through Instructional Technology in Math and Science*, Hershey, PA, IGI Global, pp. 1–20, 2012.

[SAD 12] SADEH I., ZION M., "Which type of inquiry project do high school biology students prefer: open or guided?", *Research in Science Education*, vol. 42, no. 5, pp. 831–848, 2012.

[SAD 09] SADLER T.D., "Situated learning in science education: socio-scientific issues as contexts for practice", *Studies in Science Education*, vol. 45, no. 1, pp. 1–42, 2009.

[SAL 79] SALOMON G., *Interaction of Media, Cognition, and Learning*, San Francisco: Jossey-Bass, 1979.

[SAL 87] SALOMON G., PERKINS D.N., "Transfer of cognitive skills from programming: When and how?", *Journal of Educational Computing Research*, vol. 3, pp. 149–169, 1987.

[SAL 98] SALOMON G., PERKINS D.N., "Individual and social aspects of learning", *Review of Research in Education*, vol. 23, pp. 1–24, 1998.

[SAL 91] SALOMON G., PERKINS D., GLOBERSON T., "Partners in cognition: Extending human intelligence with intelligent technologies", *Educational Researcher*, vol. 20, no. 4, pp. 2–9, 1991.

[SAL 96] SALOMON G., PERKINS D.N., "Learning in wonderland: What computers really offer education", in KERR S. (ed.), *Technology and the Future of Education*, Chicago: University of Chicago Press, pp. 111–130, 1996.

[SAL 10] SALVINI P., LASCHI C., DARIO P., "Design for acceptance: improving robots' coexistence in human society", *International Journal of Social Robotics*, vol. 2, pp. 451–460, 2010.

[SAY 10] SAYLOR M., SOMANADER M., LEVIN, D. *et al.*, "How do young children deal with hybrids of living and non-living things: The case of humanoid robots", *British Journal of Developmental Psychology*, vol. 28, pp. 835–851, 2010.

[SCH 99] SCHANK R.C., BERMAN T.R., MACPHERSON K.A., "Learning by doing", in REIGELUTH C.M. (ed.), *Instruction Design Theories and Models*, Hillsdale: Erlbaum Associates, 1999.

[SCH 05] SCHWARZ C.V., WHITE B.Y., "Metamodeling knowledge: Developing students' understanding of scientific modeling", *Cognition and Instruction*, vol. 23, no. 2, pp. 165–205, 2005.

[SCH 13] SCHAEFER K.E., The perception and measurement of human-robot trust, PhD Thesis, University of Central Florida, 2013.

[SEV 10] SEVERSON R.L., CARLSON S.M., "Behaving as or behaving as if? Children's conceptions of personified robots and the emergence of a new ontological category", *Neural Networks, Special Issue on Social Cognition: From Babies to Robots*, vol. 23, pp. 1099–1103, 2010.

[SFA 91] SFARD A., "On the dual nature of mathematical conceptions: reflections on processes and objects as different sides of the same coin", *Educational Studies in Mathematics*, vol. 22, no. 1, pp. 1–36, 1991.

[SHA 09] SHAW-GARLOCK G., "Looking forward to sociable robots", *International Journal of Social Robotics*, vol. 1, pp. 249–260, 2009.

[SHA 99] SHANK R.C., BERMAN T.R., MACPHERSON K.A., "Learning by doing", in REIGELUTH C.M. (ed.), *Instructional-Design Theories and Models: A New Paradigm of Instructional Theory*, Mahwah: Lawrence Erlbaum Associates, vol. II, pp. 161–81, 1999.

[SHI 07] SHIN N., KIM S., "Learning about, from, and with robots: students' perspectives", *16th IEEE International Conference on Robot & Human Interactive Communication (RO-MAN)*, Jeju, Korea, 2007.

[SHI 03] SHINOZAWA K., REEVES B., WISE K. *et al.*, "Robots as new media: a cross-cultural examination of social and cognitive responses to robotic and on-screen agents", *Proceedings of the International Communication Association*, San Diego, California, USA, pp. 998–1002, 2003.

[SHI 05] SHINOZAWA K., NAYA F., YAMATO J. *et al.*, "Differences in effect of robot and screen agent recommendations on human decision-making", *International Journal of Human-Computer Studies*, vol. 62, no. 2, pp. 267–279, 2005.

[SHO 89] SHORT E.J., WEISSBERG-BENCHELL J.A., "The triple alliance for learning: cognition, metacognition, and motivation", in MCCORMICK C.B., MILLER G.E., PRESSLEY M. (eds), *Cognitive Strategy Research: From Basic Research to Educational Application*, New York: Springer-Verlag, pp. 33–63, 1989.

[SKL 03] SKLAR E., EGUCHI A., JOHNSON J., "RoboCupJunior: learning with educational robotics", *AI Magazine*, vol. 24, no. 2, pp. 43–47, 2003.

[SLA 11] SLANGEN L., VAN KEULEN H., GRAVEMEIJER K., "What pupils can learn from working with robotic direct manipulation environments", *International Journal of Technology and Design Education*, vol. 21, pp. 449–469, 2011.

[SLO 03] SLOMAN S.A., MALT B.C., "Artifacts are not ascribed essences, nor are they treated as belonging to kinds", *Language and Cognitive Processes*, vol. 18, nos. 5–6, pp. 563–582, 2003.

[SLO 04] SLOTTA J.D., "The web-based inquiry science environment (WISE): scaffolding knowledge integration in the science classroom", in LINN M.C., BELL P., DAVIS E. (eds), *Internet Environments for Science Education*, Lawrence Erlbaum & Associates, pp. 203–232, 2004.

[SMI 07] SMITH T.J., "The ergonomics of learning: educational design and learning performance", *Ergonomics*, vol. 50, no. 10, pp. 1530–1546, 2007.

[SOL 68] SOLLERS P., *Logiques*, Paris: Le Seuil, 1968.

[SOA 11] SOARES F., RIBEIRO F., LOPES G. *et al.*, "CK-12, university students and robots: an early start", *Proceedings of IEEE EDUCON Education Engineering – Learning Environments and Ecosystems in Engineering Education*, Amman, Jordan, 2011.

[SOU 04] SOUGNÉ J., "Les raisonnements temporels en robotique pédagogique: deux stratégies cognitive de synchronization", in DENIS B., BARON G.L. (eds), *Regards sur la Robotique pédagogique*, Actes du quatrième colloque sur la robotique pédagogique, Liège: Paris, INRP, pp. 85–91, 2004.

[STE 06] STEINFELD A., FONG T., KABER D. *et al.*, "Common metrics for human-robot interaction", *Proceedings of the 1st ACM/IEEE International Conference on Human Robot Interaction*, New York, NY, ACM, pp. 33–40, 2006.

[STE 09] STERNBERG R.J., STERNBERG K., *Cognitive Psychology* (6th Edition), Belmont: Wadsworth, Cengage Learning, 2009.

[STR 92] STROMMEN E.F., LINCOLN B., "Constructivism, technology, and the future of classroom learning", *Education and Urban Society*, vol. 24, pp. 466–476, 1992.

[SUL 08] SULLIVAN F.V., "Robotics and science literacy: thinking skills, science process skills and systems understanding", *Journal of Research in Science Teaching*, vol. 45, no. 3, pp. 373–394, 2008.

[SUL 15a] SULLIVAN A., BERS M.U., "Robotics in the early childhood classroom: Learning outcomes from an 8-week robotics curriculum in pre-kindergarten through second grade", *International Journal of Technology and Design Education*, vol. 26, no. 1, pp. 3–20, 2015.

[SUL 15b] SULLIVAN A., ELKIN M., BERS M.U., "KIBO robot demo: engaging young children in programming and engineering", *Proceedings of the 14th International Conference on Interaction Design and Children*, ACM, Boston, MA, USA, pp. 418–421, 2015.

[SUN 06] SUN H., ZHANG P., "The role of moderating factors in user technology acceptance", *International Journal of Human-Computer Studies*, vol. 64, no. 2, pp. 53–78, 2006.

[SYR 07] SYRDAL D.S., KOAY K.L., WALTERS M.L. *et al.*, "A personalized robot companion? – The role of individual differences on spatial preferences in HRI scenarios", *Proceedings of the 16th IEEE International Symposium on Robot and Human interactive Communication*, Jeju, Korea, pp. 1143–1148, 2007.

[TAN 05] TANAKA F., FORTENBERRY B., AISAKA K. *et al.*, "Developing dance interaction between QRIO and toddlers in a classroom environment: plans for the first steps", *Proceedings of the IEEE International Workshop on Robots and Human Interactive Communication*, Nashville, USA, pp. 223–228, 2005.

[TEJ 06] TEJADA S., TRAFT N., HUTSON M. *et al.*, "Educational robots: three models for the research of learning theories and human-robot interaction", *Proceedings of the AAAI Robotics Workshop*, Boston, Massachusetts, pp. 70–76, 2006.

[THO 00] THOMAS J.W., A review of research on project-based learning, San Rafael, CA: Autodesk, available at http://www.k12reform.org/foundation/pbl/research, 2000.

[THR 04] THRUN S., "Towards a framework for human-robot interaction", *Human Computer Interaction*, vol. 19, nos. 1–2, pp. 9–24, 2004.

[TIJ 06] TIJUS C., POITRENAUD S., BOUCHON-MEUNIER B. *et al.*, "Le cartable électronique: sémantique de l'utilisabilité et aide aux apprentissages", *Psychologie Française*, vol. 51, no. 1, pp. 87–101, 2006.

[TRE 97] TRETTEN R., ZACHARIOU P., *Learning about Project-based Learning: Assessment of Project-based Learning in Tinkertech Schools*, San Rafael: Autodesk, 1997.

[TRO 05] TROUCHE L., "Construction et conduite des instruments dans des apprentissages mathématiques: Nécessité des orchestrations", *Recherches en Didactique des Mathématiques*, vol. 25, no. 1, pp. 91–138, 2005.

[TUR 95] TURKLE S., *Life on the Screen: Identity in the Age of the Internet*, New York, Simon and Schuster, 1995.

[TUR 05] TURKLE S., *The Second Self: Computers and the Human Spirit*, Cambridge: The MIT Press, 2005.

[TUR 11] TURKLE S., *Alone Together. Why we Expect more from Technology and Less from Each Other*, New York: Basic Books, 2011.

[VAN 14a] VAN DEN BRULE R., DOTSCH R., BIJLSTRA G. *et al.*, "Do robot performance and behavioral style affect human trust?", *International Journal of Social Robotics*, vol. 6, no. 4, pp. 519–531, 2014.

[VAN 14b] VAN DE KEERE K., MESTDAGH N., "An European dimension to Inquiry: the development and evaluation of the Pri-Sci-Net inquiry activities", in GATT S., MESTDAGH N., VAN DE KEERE K. *et al.* (eds), *The Pri-Sci-Net Project: Bringing Inquiry at Primary School*, 2014.

[VAN 96] VAN DUUREN M., SCAIFE M., "A robot's brain controls itself, because a brain hasn't got a brain, so it just controls itself – young children's attributions of brain related behaviour to intelligent artefacts", *European Journal of Psychology of Education*, vol. 11, pp. 365–376, 1996.

[VAN 10] VAN ECK R., *Interdisciplinary Models and Tools for Serious Games: Emerging Concepts and Future Directions*, Hershey, PA, IGI Global, 2010.

[VAN 91] VAN HORN R., *Advanced Technology in Education*, Pacific Grove: Brooks/Cole, 1991.

[VAN 07] VAN LITH P., "Teaching robotics in primary and secondary schools", *Proceedings ComLab International Conference Computerised Laboratory in Science and Technology Education*, Radovljica, Slovenia, p. 23, 2007.

[VEN 00] VENEZKY R.L., "The digital divide within formal school education: causes and consequences", Schooling for Tomorrow. Learning to Bridge the Digital Divide, OECD and University of Delaware, United States, 2000.

[VEN 03] VENKATESH V., MORRIS M.G., DAVIS G.B. *et al.*, "User acceptance of information technology: Toward a unified view", *MIS Quarterly*, vol. 27, no. 3, pp. 425–478, 2003.

[VEN 11] VEENSTRA D.N., EVERS V., "The development of an online research tool to investigate children's social bonds with robots", *Human-Robot Personal Relationships. Series Lecture Notes of the Institute for Computer Sciences, Social Informatics and Telecommunications Engineering*, vol. 59, pp. 19–26, 2011.

[VER 96] VERMUNT J.D., "Metacognitive, cognitive, and affective aspects of learning styles and strategies", *Higher Education*, vol. 31, pp. 25–50, 1996.

[VID 14] VIDAL D., GAUSSIER P., "Un robot comme personne, ontologies comparée et expérimentale au musée du quai Branly", *Terrain*, vol. 62, pp. 152–165, 2014.

[VYG 62] VYGOTSKY L.S., *Thought and Language*, Cambridge: The MIT Press, 1962.

[VYG 78] VYGOTSKY L.S., *Mind in Society: The Development of Higher Psychological Processes*, Cambridge: Harvard University Press, 1978.

[VOS 01] VOSNIADOU S., "How children learn", Educational Practices Series 7, The International Academy of Education (IAE) and the International Bureau of Education (UNESCO), 2001.

[WAL 05] WALTERS M.L., DAUTENHAHN K., BOEKHORST R. *et al.*, "The influence of subjects' personality traits on personal spatial zones in a human-robot interaction experiment", *Proceedings of IEEE International Workshop on Robot and Human Interactive Communication*, Nashville, USA, pp. 347–352, 2005.

[WAN 10] WANG Y., WIDROW B., ZHANG B. *et al.*, Perspectives on cognitive informatics and its future development: summary of Plenary Panel II of IEEE ICCI'10, IEE Computer Society, Los Alamitos, CA, USA, 2010.

[WEI 08] WEISS A., BERNHAUPT R., LANKES, "The influence of control on the acceptance of ambient intelligence by elderly people: An explorative study", in AARTS EMILE H.L., CROWLEY JAMES L. *et al.* (eds), *Ambient Intelligence – European Conference*, Nuremberg, Germany, pp. 58–74, 2008.

[WEI 09] WEISS A., BERNHAUPT R., LANKES M. *et al.*, "The USUS evaluation framework for human-robot interaction", in *Proceedings of the Symposium on New Frontiers in Human-Robot Interaction*, Edinburgh, Scotland, UK, pp. 158–165, 2009.

[WEL 10] WELCH K.C., LAHIRI U., WARREN Z. *et al.*, "An approach to the design of socially acceptable robots for children with autism spectrum disorders", *International Journal of Social Robotics*, vol. 2, no. 4, pp. 391–403, 2010.

[WEN 09] WENG Y.H., CHEN C.H., SUN C.T., "Toward the human–robot co-existence society: on safety intelligence for next generation robots", *International Journal of Social Robotics*, vol. 1, pp. 267–282, 2009.

[WHI 98] WHITE B., FREDERICKSEN J., "Inquiry, modeling, and metacognition: making science accessible to all students", *Cognition and Instruction*, vol. 16, pp. 3–118, 1998.

[WHE 00] WHEELER G., "The three faces of inquiry", in MINSTRELL J., VAN ZEE E. (eds), *Inquiring into Inquiry Learning and Teaching in Science*, Washington DC: AAAS, pp. 14–19, 2000.

[WHI 07] WHITTIER L.E., ROBINSON M., "Teaching evolution to non-english proficient students by using lego robotics", *American Secondary Education*, vol. 35, no. 3, pp. 19–28, 2007.

[WIG 98] WIGGINS G., MCTIGHE J., Understanding by design, Association for Supervision and Curriculum Development, Alexandria, VA, 1998.

[WIL 07] WILLIAMS D., MA Y., PREJEAN L. *et al.*, "Acquisition of physics content knowledge and scientific inquiry skills in a robotics summer camp", *Journal of Research on Technology in Education*, vol. 40, no. 2, pp. 201–216, 2007.

[WIL 85] WILSON S.W., "Knowledge growth in an artificial animal", in GREFENSTETTE J.J. (ed.), *Proceedings of the First International Conference on Genetic Algorithms and Their Applications*, Hillsdale: Lawrence Erlbaum Associates, pp. 16–23, 1985.

[WIN 04] WINDSCHITL M., "Caught in the cycle of reproducing folk theories of "Inquiry": how pre-service teachers continue the discourse and practices of an atheoretical scientific method", *Journal of Research in Science Teaching*, vol. 41, no. 5, pp. 481–512, 2004.

[WIN 08] WING J.M., "Computational thinking and thinking about computing", *Philosophical Transactions of the Royal Society A*, vol. 366, pp. 3717–3725, 2008.

[WOO 76] WOOD D.J., BRUNER J.S., ROSS G., "The role of tutoring in problem solving", *Journal of Child Psychiatry and Psychology*, vol. 17, no. 2, pp. 89–100, 1976.

[WOU 09] WOUTERS P., VAN DER SPEK E.D., VAN OOSTENDORP H., "Current practices in serious game research: A review from a learning outcomes perspective", in CONNOLLY T.M., STANSFIELD M., BOYLE L. (eds), *Games-Based Learning Advancements for Multisensory Human Computer Interfaces: Techniques and Effective Practices*, Hershey: IGI Global, pp. 232–250, 2009.

[WUL 14] WULLENKORD R., EYSSEL F., "Improving attitudes towards social robots using imagined contact", *Proceedings of the 23rd IEEE International Symposium on Robot and Human Interactive Communication* (*RO-MAN*), Edinburgh, Scotland, UK, pp. 489–49, 2014.

[WYE 04] WYETH P., VENZ M., WYETH G., "Scaffolding children's robot building and programming activities", in POLANI D., BROWNING B., BONARINI A. *et al.* (eds), *RoboCup 2003: Robot Soccer World Cup VII, Lecture Notes in Computer Science*, vol. 3020, pp. 308–319, 2004.

[WU 14] WU X., ZHU X., WU G.-Q. *et al.*, "Data mining with big data", *IEEE Transactions on, Knowledge and Data Engineering*, vol. 26, no. 1, pp. 97–107, 2014.

[YAG 11] YAGODA R.E., What! You want me to trust a robot? The Development of a Human Robot Interaction (HRI) Trust Scale, Master's thesis, North Carolina State University, Raleigh, North Carolina, 2011.

[YAN 02] YANCO H.A., Drury J.L., "A taxonomy for human-robot interaction." *Proceedings of the AAI Fall Symposium on Human-Robot Interaction.* AAAI Technical Report FS-02-03, Falmouth, MA, pp. 111–119, 2002.

[YOU 09] YOUNG J.E., HAWKINS R., SHARLIN E. *et al.*, "Toward acceptable domestic robots: applying insights from social psychology", *International Journal of Social Robotics*, vol. 1, pp. 95–108, 2009.

[YU 15] YU A.J., "Decision-making tasks", *Encyclopedia of Computational Neuroscience*, pp. 931–937, 2015.

[ZAA 08] ZAAD L., BEN ALLOUCH S., "The influence of control on the acceptance of ambient intelligence by elderly people: an explorative study", *Ambient Intelligence, Series Lecture Notes in Computer Science*, vol. 5355, pp. 58–74, 2008.

[ZAJ 65] ZAJONC R.B., "Social facilitation", *Science*, vol. 149, pp. 269–274, 1965.

[ZAJ 80] ZAJONC R.B., "Compresence psychology of group influence", in PAULUS P.B. (ed.), *Psychology of Group Influence*, pp. 35–60, 1980.

[ZER 14] ZERAFA I., GATT S., "Implementing a science curriculum reflecting an inquiry based approach in the upper primary years", *IPSE Journal*, vol. 1, no. 2, pp. 13–26, 2014.

[ZHA 10] ZHANG G., "Virtual robots and its education application", *Proceedings of Educational Technology and Computer Science Second International Workshop*, vol. 1, pp. 793–796, 2010.

[ZIB 11] ZIBETTI E., CHEVALIER A., EYRAUD R., "What type of information displayed on digital scheduling software facilitates reflective planning tasks for students? Contributions to the design of a school task management tool", *Computers in Human Behavior*, vol. 28, pp. 591–607, 2011.

[ZIM 89] ZIMMERMAN B.J., "Models of self-regulated learning and academic achievement", in ZIMMERMAN B.J., SCHUNK D.H. (eds), *Self-Regulated Learning and Academic Achievement: Theory, Research, and Practice*, New York: Springer, pp. 1–25, 1989.

[ZIM 90] ZIMMERMAN B.J., "Self-regulated learning and academic achievement: an overview", *Educational Psychologist*, vol. 25, no. 1, pp. 3–17, 1990.

[ZIO 05] ZION M., SLEZAK M., "It takes two to tango: In dynamic inquiry, the self-directed student acts in association with the facilitating teacher", *Teaching and Teacher Education*, vol. 21, no. 7, pp. 875–894, 2005.

[ZIO 12] ZION M., MENDELOVICI R., "Moving from structured to open inquiry: challange and limits", *Science Education International*, vol. 23, no. 4, pp. 383–399, 2012.

# Index

## A, C



## D, E, F



## G, H, I

## L, M, N



## O, P, R



## S, T, W

Other titles from

iSTE

in

Information Systems, Web and Pervasive Computing

## 2016

BEN CHOUIKHA Mona
*Organizational Design for Knowledge Management*

BÉRANGER Jérôme
*Big Data and Ethics: The Medical Datasphere*

BERTOLO David
*Interactions on Digital Tablets in the Context of 3D Geometry Learning (Human-Machine Interaction Set – Volume 2)*

EL FALLAH SEGHROUCHNI Amal, ISHIKAWA Fuyuki, HÉRAULT Laurent, TOKUDA Hideyuki
*Enablers for Smart Cities*

KITAJIMA Munéo
*Memory and Action Selection in Human–Machine Interaction (Human–Machine Interaction Set – Volume 1)*

LAGRAÑA Fernando
*E-mail and Behavioral Changes: Uses and Misuses of Electronic Communications*

MONINO Jean-Louis, SEDKAOUI Soraya
*Big Data, Open Data and Data Development*

SALGUES Bruno
*Health Industrialization*

VENTRE Daniel
*Information Warfare – 2nd edition*

## 2015

ARDUIN Pierre-Emmanuel, GRUNDSTEIN Michel,
ROSENTHAL-SABROUX Camille
*Information and Knowledge System*
*(Advances in Information Systems Set – Volume 2)*

BÉRANGER Jérôme
*Medical Information Systems Ethics*

BRONNER Gérald
*Belief and Misbelief Asymmetry on the Internet*

IAFRATE Fernando
*From Big Data to Smart Data*
*(Advances in Information Systems Set – Volume 1)*

KRICHEN Saoussen, BEN JOUIDA Sihem
*Supply Chain Management and its Applications in Computer Science*

NEGRE Elsa
*Information and Recommender Systems*
*(Advances in Information Systems Set – Volume 4)*

POMEROL Jean-Charles, EPELBOIN Yves, THOURY Claire
*MOOCs*

SALLES Maryse
*Decision-Making and the Information System (Advances in Information Systems Set – Volume 3)*

SAMARA Tarek
*ERP and Information Systems: Integration or Disintegration (Advances in Information Systems Set – Volume 5)*

## 2014

DINET Jérôme
*Information Retrieval in Digital Environments*

HÉNO Raphaële, CHANDELIER Laure
*3D Modeling of Buildings: Outstanding Sites*

KEMBELLEC Gérald, CHARTRON Ghislaine, SALEH Imad
*Recommender Systems*

MATHIAN Hélène, SANDERS Lena
*Spatio-temporal Approaches: Geographic Objects and Change Process*

PLANTIN Jean-Christophe
*Participatory Mapping*

VENTRE Daniel
*Chinese Cybersecurity and Defense*

## 2013

BERNIK Igor
*Cybercrime and Cyberwarfare*

CAPET Philippe, DELAVALLADE Thomas
*Information Evaluation*

LEBRATY Jean-Fabrice, LOBRE-LEBRATY Katia
*Crowdsourcing: One Step Beyond*

SALLABERRY Christian
*Geographical Information Retrieval in Textual Corpora*

## 2012

BUCHER Bénédicte, LE BER Florence
*Innovative Software Development in GIS*

GAUSSIER Eric, YVON François
*Textual Information Access*

STOCKINGER Peter
*Audiovisual Archives: Digital Text and Discourse Analysis*

VENTRE Daniel
*Cyber Conflict*

## 2011

BANOS Arnaud, THÉVENIN Thomas
*Geographical Information and Urban Transport Systems*

DAUPHINÉ André
*Fractal Geography*

LEMBERGER Pirmin, MOREL Mederic
*Managing Complexity of Information Systems*

STOCKINGER Peter
*Introduction to Audiovisual Archives*

STOCKINGER Peter
*Digital Audiovisual Archives*

VENTRE Daniel
*Cyberwar and Information Warfare*

## 2010

BONNET Pierre
*Enterprise Data Governance*

BRUNET Roger
*Sustainable Geography*

CARREGA Pierre
*Geographical Information and Climatology*

CAUVIN Colette, ESCOBAR Francisco, SERRADJ Aziz
*Thematic Cartography – 3-volume series*
*Thematic Cartography and Transformations – volume 1*
*Cartography and the Impact of the Quantitative Revolution – volume 2*
*New Approaches in Thematic Cartography – volume 3*

LANGLOIS Patrice
*Simulation of Complex Systems in GIS*

MATHIS Philippe
*Graphs and Networks – 2nd edition*

THERIAULT Marius, DES ROSIERS François
*Modeling Urban Dynamics*

## 2009

BONNET Pierre, DETAVERNIER Jean-Michel, VAUQUIER Dominique
*Sustainable IT Architecture: the Progressive Way of Overhauling Information Systems with SOA*

PAPY Fabrice
*Information Science*

RIVARD François, ABOU HARB Georges, MERET Philippe
*The Transverse Information System*

ROCHE Stéphane, CARON Claude
*Organizational Facets of GIS*

## 2008

BRUGNOT Gérard
*Spatial Management of Risks*

FINKE Gerd
*Operations Research and Networks*

GUERMOND Yves
*Modeling Process in Geography*

KANEVSKI Michael
*Advanced Mapping of Environmental Data*

MANOUVRIER Bernard, LAURENT Ménard
*Application Integration: EAI, B2B, BPM and SOA*

PAPY Fabrice
*Digital Libraries*

## 2007

DOBESCH Hartwig, DUMOLARD Pierre, DYRAS Izabela
*Spatial Interpolation for Climate Data*

SANDERS Lena
*Models in Spatial Analysis*

## 2006

CLIQUET Gérard
*Geomarketing*

CORNIOU Jean-Pierre
*Looking Back and Going Forward in IT*

DEVILLERS Rodolphe, JEANSOULIN Robert
*Fundamentals of Spatial Data Quality*

Printed and bound by CPI Group (UK) Ltd, Croydon, CR0 4YY
15/09/2026
14962342-0002